ORDER AND DISORDER IN MATTER

LECTURE NOTES AND SUPPLEMENTS IN PHYSICS
John David Jackson and David Pines, *Editors* (Nos. 1–13)

Volumes of the Series published from 1962–1974 are not officially numbered. The parenthetical numbers shown are designed to aid librarians and bibliographers to check the completeness of their holdings.

(1)	John David Jackson	Mathematics for Quantum Mechanics: An Introductory Survey of Operators, Eigenvalues, and Linear Vector Spaces, 1962
(2)	Willem Brouwer	Matrix Methods in Optical Instrument Design, 1964
(3)	R. Hagedorn	Relativistic Kinematics: A Guide to the Kinematic Problems of High-Energy Physics, 1964 (5th printing, 1980)
(4)	Robert S. Knox and Albert Gold	Symmetry in the Solid State, 1964
(5)	David Pines	Elementary Excitations in Solids: Lectures on Phonons, Electrons, and Plasmons, 1963 (3rd printing, 1978)
(6)	Gabriel Barton	Introduction to Dispersion Techniques in Field Theory, 1965
(7)	David Bohm	The Special Theory of Relativity, 1965 (2nd printing, 1979)
(8)	David Park	Introduction to Strong Interactions: A Lecture-Note Volume, 1966
(9)	Hans A. Bethe and Roman W. Jackiw	Intermediate Quantum Mechanics, 1968 (5th printing, 1982)
(10)	Gordon Baym	Lectures on Quantum Mechanics, 1969 (9th printing, with corrections, 1981)
(11)	K. Nishijima	Fields and Particles: Field Theory and Dispersion Relations, 1969 (4th printing, 1980)
(12)	B. H. Bransden	Atomic Collision Theory, 1970
(13)	R. D. Sard	Relativistic Mechanics: Special Relativity and Classical Particle Dynamics, 1970

LECTURE NOTES AND SUPPLEMENTS IN PHYSICS
David Pines, *Editor*

Volumes published from 1975 onward are being numbered as an integral part of the bibliography:

Number		
14	Hans Frauenfelder and Ernest M. Henley	Nuclear and Particle Physics, A: Background and Symmetries, 1975
15	H. W. Wyld	Mathematical Methods for Physics, 1976 (3rd printing, 1981)
16	Werner O. Amrein, Josef M. Jauch, and Kalyan B. Sinha	Scattering Theory in Quantum Mechanics: Physical Principles and Mathematical Methods, 1977
17	B. H. Bransden	Atomic Collision Theory, Second Edition, 1983
18	Giorgio Careri	Order and Disorder in Matter, 1984

ORDER AND DISORDER IN MATTER

Giorgio Careri

University of Rome

Rome, Italy

English-language translation by

Kristin Jarratt

1984

The Benjamin/Cummings Publishing Company, Inc.
Advanced Book Program
Menlo Park, California

Reading, Massachusetts • London • Amsterdam • Don Mills, Ontario • Sydney

Sponsoring Editors	Richard W. Mixter, Carl W. Harris
Production	Jan deProsse
Interior Design	Hannus Design Associates
Cover Design	Ben Branscom

Library of Congress Cataloging in Publication Data

Careri, Giorgio.
Order and disorder in matter.

Translation of: Ordine e disordine nella materia.
Bibliography: p.
Includes index.
1. Order-disorder models. 2. Matter. I. Title.
QC173.4.073C3713 1984 530 84-384
ISBN 0-8053-1700-7 (Hardbound)
ISBN 0-8053-1725-2 (Paperbound)

ABCDEFGHIJ-AL-8987654

The Benjamin/Cummings Publishing Company, Inc.
2727 Sand Hill Road
Menlo Park, CA 94025

CONTENTS

PREFACE

Audience and Approach to the Subject

I have written this book for both undergraduate and graduate students of science and for scholars interested in the synthetic description of material systems that emerges from the general concepts of order and disorder. My concern here is to cover the macroscopic description of matter that is typical of statistical physics and, to some extent, other fundamental interdisciplinary aspects such as cosmic order and biological order. The problems in this field originate from the widely observed fact that matter takes on macroscopic ordered structures that cannot generally be predicted by direct extrapolation from microscopic properties. Historical developments have shown that these structures can be analyzed only when we progressively refine our concept of order. This point is of broad interest in modern scientific thought, because it allows us to extend the use of our concept of order to the analysis of a wider, natural scenario not restricted to physics alone.

As I have said, one general feature of the state of matter we are considering is that, as the complexity of the system increases, there is a parallel emergence of a phenomenology that cannot easily be extrapolated from small-scale properties. For instance, if we start with only the notion of the dynamic laws governing movement of single molecules in a gas, it would be difficult to predict the existence of a general phenomenon such as sound, which is caused by a succession of compression and rarefaction zones themselves interrelated by such simple regularities as those of wave motion. Likewise, who would ever think that a crystal, with its typical spatial symmetries, would be formed when we lower the temperature of a homogeneous and isotropic liquid composed of perfectly spherical atoms?

These two examples from everyday life share the feature of long-range order expressed in the first case as the coherence of the sound wave and in the second case as the lattice structure of the crystal. As we shall see in this book, nature presents us with numerous spontaneous or induced phenomena of order rarely predictable from a knowledge of the microscopic system. Thus in this area a phenomenological approach is now perfectly appropriate because no general microscopic theory can predict the macroscopic structure of any material system under any boundary conditions. Since these structures are experimentally observed in systems of some complexity we must perform a case-by-case study; and now that this process has been extended to systems of higher complexity, it is leading to a wider conception of our idea of order.

The level of complexity of the material system considered in this book is on a par with that normally used in the study of atomic physics, solid-state physics, and chemistry, where in principle all phenomenology can be traced back to the laws of nonrelativistic quantum mechanics and statistical mechanics. In simple terms, it can be said that our material system is a large assembly of atomic nuclei and electrons, and thus the energy of the system can be formally written as the sum of the kinetic energies of the nuclei and electrons and of the potential energy that contains all the electrical and magnetic interactions of the system, whether internal or external. It must be mentioned that none of the typical concepts of chemistry (i.e., valence and bond strength) or of solid-state physics (i.e., lattice structure and energy bands) can emerge from the sole consideration of this interpretation of the system's total energy. We know that these concepts emerge only after approximation—that is, following the introduction of a proper model that will let the meaning of a particular concept emerge. The same is true for the concepts of order and disorder, which are among the more general concepts that bridge the gap between the macroscopic and microscopic worlds and are quite effective in describing the material systems at this level of complexity. Consequently, to enlarge our view of these concepts in this book, we shall formulate more and more elaborate models and thus make them applicable to conditions of ever-increasing complexity.

To achieve this progressive enlargement of the concept of order in a book of this size, I have limited my choices to phenomena that are most significant at present. To avoid subjects that inevitably involve relativity and subatomic physics, I have had to omit the formation of the ordered structures of the universe, a subject that is now fully developed. In addition, I have not considered obvious ordered or disordered systems, such as crystals and gases, nor systems that are too complex, such as ordinary liquids and crystal defects. Of course, many of these structures have been discovered and immediately understood in terms of fairly simple models, with no need for important new concepts. On the other hand, we have taken half a century to understand the apparent strange behavior of superfluids; but our knowledge of the order hidden in the superfluid state has changed our view of order in matter and has prepared the way for the general theory of critical phenomena developed in the 1970s. A parallel conceptual development also took place recently in systems outside thermodynamic equilibrium, in the hope of including biological systems in this area as well. Therefore, this seemed to me the best time to show—I hope clearly and concisely—how the concepts of order and disorder have been evolving to contain an ever wider phenomenology, ranging from superfluids to biomolecules. I also have personal reasons for making this choice, involving my experimental work on superfluids in this conclusive chapter on physics and my present involvement with systems of biological interest in the attempt to interpret them as ordered physical structures. Naturally, the process of fitting these concepts into reality in ever more complex physical systems is already under way; one of the objectives of this book is to point out the limits of our present knowledge and thus indicate problems still open to research.

Organization of the Book

I have divided this book into five chapters; these should be seen as the most significant points in a single line of thought. The first chapter is conceived of as an area of contact between what readers probably know about general physics and the concepts that will be developed in the rest of the book. Instead of briefly summarizing statistical thermodynamics, I felt that it would be wiser to recall several models to which statistical thermodynamics can be applied, pointing out the difficulty of treating the real system behind each model in this way. Thus the first chapter plays a pedagogical role and can easily be skipped by readers who are already familiar with these concepts; however, this chapter can be used independently from the following chapters by students taking beginning courses in general physics and structure of matter. Within the elementary context of statistical thermodynamics considered in this chapter, order and disorder appear to be two well-defined entities that merely complement each other. This position is drastically altered by the study of superfluids in the second chapter, because here order and disorder become two different ingredients in the description of the peculiar quantomechanical nature of the system.

The study of superfluids is carried out almost in an historical vein, to mark the development of our ideas about superfluid helium in this simple and real system that has played an historical role in many-body quantum systems similar to that of the hydrogen atom in few-body quantum systems. A rapid summary of the most important experimental facts allows for self-contained and elementary treatments of the problem and comprehension of the important new conception reached toward mid-century, which sees order as a coherence of the fundamental collective state and disorder as a gas of quasi-particles distributed over excited collective states.

In the third chapter, on the basis of the particular and remarkable case of superfluids, I have outlined the more general question regarding the emergence of an order parameter as symmetry breaking in a material system. I have started from common phenomenology presented by a vast class of phase transitions in conditions of thermodynamic equilibrium, but I have had to develop the notion of order parameter in more abstract form: Without appropriate mathematical language, I might have had to sacrifice the clarity that can otherwise be achieved in this field, thanks to the most recent theoretical developments.

The phenomenology of phase transitions at equilibrium is the obvious reference point for the disorder–order transitions outside equilibrium. This is discussed in the fourth chapter. This subject is still in the assessment stage, and it is also of interest here to show the new conceptual change that is now required to understand the phenomenology of the laser, among other things. The laser is the prototype of an open physical system, and it contains all necessary features for the statement of the induced order in the new terms of correlation among events. We need this new conceptual outlook to discuss the functional order of biology in the fifth chapter; this chapter is essentially different from the rest of the book because here I have tried to deal in an original manner with a subject that is plainly interdisciplinary and is seen as the extrapolation of the physical line of thought formed in the previous chapters. The need to avoid biochemical details has made this discussion even more qualitative here, so I have had to limit it to a few points of physical interest alone. Finally, I have added an appendix containing a few elementary considerations on two very general problems, cosmic order and the very idea of order in science; these topics border upon the more specific theme of the book.

It can be seen that the conception of order in matter is progressively modified in the book, to fit it to increasingly complex systems under more realistic conditions. The initial point of view here is the outlook common to statistical thermodynamics, where disorder is measured by entropy and order is simply the complementary aspect. Of course, this viewpoint is helpful in only a few model systems, and the impossibility of evaluating the thermodynamic functions in a simple system such as liquid helium makes a new point of view necessary. Next, the long-range order of superfluid helium is identified by the presence of a collective wave function connected with the fundamental energy state of the quantum system. This fact is then generalized to include even more complex material aggregates through the concept of order parameter, a quantity with few internal components that controls the spontaneous transitions toward order below a critical temperature far from absolute zero. However, the discussion of critical phenomena shows the need to consider the space–time fluctuations of the order parameter as well, and the statistical character of long-range order is then established with the introduction of a coherence length; within this is a permanent statistical correlation in the distribution of matter. Likewise, through the order parameter, transitions are described in open systems outside equilibrium, and we are faced with the alternative possibility of conceiving of order as the space–time distribution correlation itself. The example of the laser clarifies this point, because here the elementary event can easily be identified in the emission of a light quantum. The space–time correlation can be viewed either among emission events or in the distribution of excited atoms, and it is clearly seen in the wave character of the coherent electromagnetic field. Finally, the correlation among events is the best way to represent the state of order of an open and complex system; consideration of a nonphysical system such as the biological organism makes this even more clear. In this case, if the elementary event is identified in the chemical transformation of matter that takes place in the cells, it is possible to describe the steady functional order as a correlation between significant events planned by the genetic code. As a consequence, *the statistical space–time correlation now seems to be the natural evolution and merging of the two concepts of order and disorder.* Statistical thermodynamics, through entropy, furnished a correct measurement of disorder that unfortunately turned out to be sterile for the more interesting real systems. Furthermore, consideration of the systems outside equilibrium leads us to shift our attention from the elements of matter to these very events, which take place in processes involving matter. We can conclude that *the closer the correlation among the events* that give rise to these processes, *the higher is the order* displayed by the open system. This is the main conclusion that I hope will emerge from this book.

This understanding of order and disorder in terms no longer static but truly dynamic—thanks to the notion of correlation among events—stems from physics but seems relevant to other fields as well. For instance, I am convinced that molecular biologists as well as many life scientists will now find it easier to express their experimental findings in terms of correlated events displayed by nature. I am also convinced that this same notion will be useful in the philosophy of science, because after Whitehead and Wittgenstein the very notion of event has reached a primary place in current philosophical thought. Thus, as often in the history of scientific ideas, the proper introduction of a new description in physics can be applied successfully to another field and later becomes embodied in the broader context of natural philosophy.

To reach a large number of students and scholars working in different fields I have been forced to eliminate a large amount of detail that would have made this book easier for the more expert students to read and, likewise, for me to write. Almost no mathematical expressions have been used. Drawings have been kept to a minimum and are presented in the form of boxes, so not to interrupt the reading of the text, with references to the bibliography for further concept development. Thus, if this book is to be used in teaching, it should be accompanied by texts with adequate quantitative development; then it could be a useful guide for stressing the ideas too often submerged by quantitative treatments. By eliminating so many details, we might be left with too few topics that are completely covered. At any rate, I hope this book will interest young people who are fascinated by these topics, as well as scholars always in search of clarification of the notions they already possess. This book is the result of lectures and classes that I have given on various occasions. I am grateful to my colleagues and students for their reactions, which I have always found enlightening.

G.C.

Rome, September 1981

CHAPTER ONE

ORDER AND DISORDER IN SOME MODEL SYSTEMS

1.1 The Perfect Molecular Gas

In this chapter we shall look at the meaning of temperature and entropy in several simple systems, to throw light on the concept of disorder from the viewpoint of statistical thermodynamics. This reconsideration is essential because of the widespread belief that heat only means energy in a disordered form and that temperature is no more than the degree of mean agitation of the molecules. Those who hold these beliefs do not realize how limited this understanding is. At the end of the eighteenth century, Rumford guessed the real meaning of heat to be a form of energy in motion; later, in 1905, Einstein offered a vivid interpretation of thermal agitation in molecular terms, with his theory of Brownian motion. But the laws of thermodynamics are much richer. In the well-known equation connecting entropy S, internal energy E, and absolute temperature T of a body in inner equilibrium,

$$\frac{dS}{dE} = \frac{1}{T},$$

we find these three fundamental quantities together in a general relationship that no longer implies any necessary reference to thermal agitation as the cause of disorder in matter. Of course, this equation holds no secrets, but to understand it fully we ought to look closer at the statistical concepts underlying these three quantities.

Given the important role of these concepts in all natural sciences, we will use several models in a less rigorous approach than that of statistical mechanics. After all, everything in physics, from Galileo's pendulum to the "perfect" gas of thermodynamics, is a continuous effort to understand more about reality by building models capable of clarifying concepts of such broad application that even the models themselves are transcended. Thus, to reach the concept of disorder in matter, we shall use a few different models that, however, are alike in that the two fundamental quantities S and T meet in them. Each of these models is of interest for its own sake and can be used for other developments, which we will only mention here. After dealing with each model as a problem in itself, we will compare them in the last section of this chapter; we will also generalize about the conclusions reached in each step.

The first model is the well-known ordinary perfect gas. This is a system of molecules in a container; these molecules are thought to be material points that obey the laws of classical mechanics without interacting forces. There can be no doubt about the importance of this model, because it gives us the first glimpse of such typical quantities of thermodynamics as pressure and temperature and enables us to calculate other quantities, such as entropy. Unfortunately, this model has one drawback: The internal energy (which is entirely kinetic) as well as the temperature and of course the entropy are all quantities that indicate molecular disorder. To be more precise, remember that elementary kinetic theory shows the mean kinetic energy for this model to be $1/2\ kT$ per degree of freedom, where k is the Boltzmann constant. This is called the *equipartition theorem of kinetic energy,* but sometimes it is erroneously called the "principle of energy equipartition." Actually, the validity of this theorem is more general, because it proves to be true even for classical systems in which energy can be expressed as the sum of quadratic terms—such as in the harmonic oscillator, where part of the energy is potential. But this theorem has led many people to the mistaken belief that temperature must always be proportional to mean disordered kinetic energy; that heat is always energy in the form of disordered motion; and that, therefore, temperature and heat are both quantities that somehow measure the same molecular disorder. In some systems energy is entirely potential and thus cannot be expressed in quadratic terms; for these systems the equipartition theorem is certainly not valid, but a thermometer will show that they display a temperature. Thus we are faced with the problem of finding out what temperature means in these cases; the best way to go about this is to consider a few specific models that exemplify these different situations. Starting from the simplest situations, already known from general physics, we will move on to more complex ones, which we shall discuss in more appropriate terms in the chapters to follow.

1.2 The Perfect Spin Gas

Let us begin by recalling the behavior of a classic magnetic dipole in an external magnetic field. If a magnetic needle is inside a medium in an external magnetic field, the needle will line up alongside the field, if it is possible to dissipate energy as heat to reach thermodynamic equilibrium. When this is

Box 1.A

THE CLASSICAL MAGNETIC DIPOLE

- Magnetic dipole of moment μ, produced by a current running through a coil

- Potential energy of a dipole in an external magnetic field H

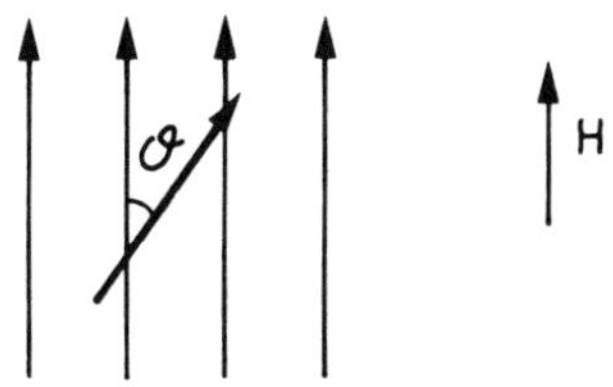

$$\mathcal{E} = \bar{\mu} \times \bar{H}$$

$$= \mu H \cos \theta$$

- Langevin curve of the mean orientation values along the field direction (k = Boltzmann's constant)

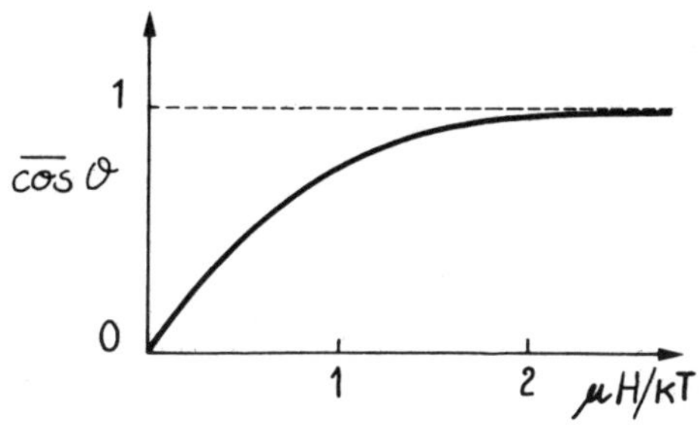

- Note that at equilibrium both order and disorder are in competition in terms of the two forms of energy

Thermal energy that creates disorder = kT
Field energy that creates order = μH

possible, we have our first example of magnetic order formation, which is expressed by the well-known Langevin curve. This curve is derived using the elementary argument of classic statistical mechanics. It offers the mean orientation of dipoles as a function of the ratio of the thermal to the magnetic energy, as shown in Box 1.A. A look at the mean behavior of this dipole as expressed by the Langevin curve shows the first evidence that *order and disorder at equilibrium are "in competition"* and that *they meet as forms of energy,* in the sense that their confrontation as quantities is in terms of energy. The Langevin curve also shows that a situation in which all the dipoles are parallel to the external field (i.e., a situation of perfect order) can only be obtained asymptotically, in the largest magnetic fields or at the lowest temperatures.

The Langevin curve holds in classical mechanics. Since we are interested in the microscopic world, we should think of this magnetic momentum as a spinning charge of atomic dimensions whose magnetic momentum is always lined up alongside the rotation axis, as required by the Ampère theorem. Today, everyone knows that the intrinsic spin of the electron, as of many elementary particles, is not zero. Now we can prove that it is possible to extend the Langevin classical theory to the case of a spin system without altering our previous conclusion regarding the competition between order and disorder in terms of energy. We can then compute that, if the magnetic moment is one Bohr magneton, as in the case of an electron, and if the external magnetic field is close to 10^4 gauss (the typical field of an electromagnet), the two resulting energies become the same only at a temperature of about one degree absolute. Thus we conclude that the spin gas must be very disordered at room temperature.

Let us now build a model for a perfect spin gas—namely, a system in which the different spins interact not with each other but only with the external field. In this respect, such a model system is analogous to the perfect molecular gas that obeys Boyle's law, because the role played by the external magnetic field in the case of the spin gas is played by pressure in the molecular gas. As we have already mentioned, Boyle's gas has the disadvantage that the internal energy, the temperature, and the entropy are all connected with disorder, and thus it is better to use the model system of quantized spins to disentangle these concepts. The development of this model requires brief mention of two general facts, space quantization and the Boltzmann distribution (see Box 1.B). Each of these facts belongs to a wide body of knowledge, the first to quantum mechanics and the second to statistical mechanics, which we shall try to use only partially here. Let it suffice to point out that space quantization means that a magnetic dipole in a magnetic field can take only certain orientations, so that its energy becomes quantized; likewise in the Bohr atom, the orbit of the electron can assume only certain orientations, and this gives rise to the magnetic quantum. Here we shall merely accept the existence of this experimental fact; indeed, space quantization was discovered in the laboratory by Stern and Gerlach even before it had been theoretically proposed. As for the Boltzmann distribution, many readers have likely come across its most particular form, the one called Maxwell's law, i.e., the distribution of molecular velocities in a perfect gas. Actually, the Boltzmann distribution is quite a general law of

Box 1.B

THE PERFECT QUANTIZED SPIN GAS

- The energy levels of one spin form a finite set because of the limited number of orientations allowed by space quantization.

$$E_i = i\mu H$$

$$\Delta E = \mu H$$

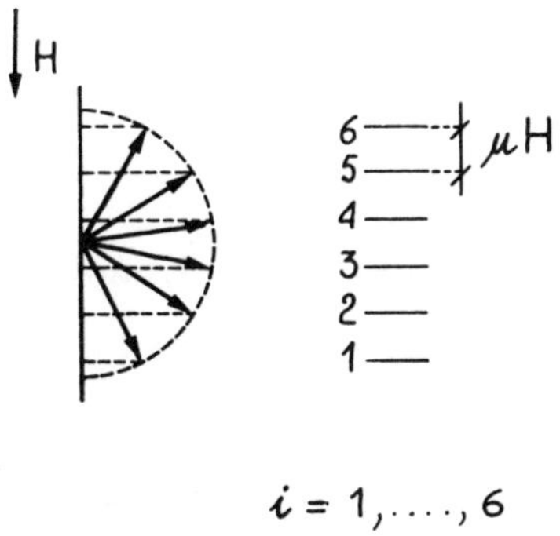

- At fixed temperature, the number of spins n_i of energy E_i drops as energy rises, according to Boltzmann's law.

$$n_i \simeq e^{-E_i/kT}$$

$$\Sigma_i n_i = N$$

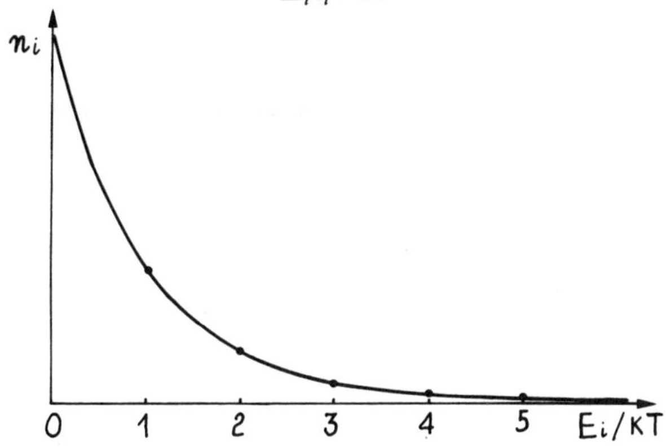

statistical physics. It states that, if a population is statistically distributed among many energy levels, then each of these levels in conditions of thermodynamic equilibrium will contain a number of individuals that will decrease exponentially with the energy of that level, as seen in Box 1.B.* Having stated these two general facts, we shall go on to use this model of perfect quantized spin gas. It should be emphasized at the outset that *the energy in this model is entirely potential,* whereas the energy in Boyle's perfect molecular gas is entirely kinetic.

1.3 Thermodynamic Quantities

First, let us try to identify the fundamental thermodynamic functions of the model we have just described, beginning with the internal energy E. By definition, this is the total energy of the system and thus, in this case, it can only be the sum of the energy of each level, multiplied by the number of individuals found there, as seen in Box 1.C. Next, we calculate the differential of the internal energy, and it seems sound definition to call the two resulting parts work W and heat Q. To verify this, let us first identify work. If we perform an operation whose only result is a variation in the external magnetic field, the levels must certainly vary because energy is the scalar product of the field strength times the magnetic momentum associated with the spin. Thus, if we increase the magnetic field, the result will be to raise all energy levels. This simply means to do work. Now, proper consideration of the other differential term will give us a direct idea of the heat involved in this process. As a matter of fact, if we perform an operation in which the energy levels are kept at constant values while the populations vary, there must be a consequent energy variation. Thus *"to change the population distribution on fixed levels" means to exchange heat with the outside.* Obviously, if this operation results in a greater population at the lower levels, heat has been removed from the system.

Again we have seen that both work and heat are forms of energy in transit between the system and the surroundings, and the internal energy is the only truly permanent property of the system. Indeed, the first law of thermodynamics shows that the internal energy is the only state function. For our model, this means that we cannot independently and arbitrarily vary the energy levels and statistical populations during a transition because, if the system is in equilibrium at a given temperature, the need to achieve Boltzmann distribution means that the variation of levels and of populations must be related. If these two operations are done separately, work or heat will be exchanged, whereas if they are done properly together, the internal

*In these and other equations, normalization factors have been omitted so that the equations will be simpler and more impressive.

Box 1.C

THERMODYNAMIC FUNCTIONS OF THE SPIN GAS

- Let us write the internal energy E as the sum of all the energies:

$$E = \sum_i E_i n_i$$

- Then differentiate:

$$dE = \sum_i (E_i \, dn_i + n_i \, dE_i) = dQ + dW$$

- We recognize that W is the work done by the external field H because

$$dE_i = i\mu \, dH$$

$$dW = \sum n_i \, dE_i$$

- Thus we identify Q as the heat, which means performing those operations that vary the populations at fixed levels:

$$dQ = \sum E_i \, dn_i$$

- Note that E is a state function, unlike W and Q, and this means that levels and populations cannot be varied separately or independently in one system.

energy will not vary from one state of equilibrium to another, because this is the only way the populations can accommodate to a new distribution when the levels change.

To explain the notion of temperature further, let us use this model by recording the population logarithm on the x-axis and the energy levels on the y-axis, as shown in Box 1.D.† This diagram gives us a simple and straightforward picture of the situation in conditions of equilibrium. The ends of the segments that measure the population logarithms must meet at one sin-

†For simplicity, we have recorded only six levels here, as for the case of a 5/2 spin.

Box 1.D

THE TEMPERATURE OF A SPIN GAS

- Let us write the Boltzmann law a different way, as

$$\ln n_i = \frac{E_i}{kT}$$

and identify temperature as the trigonometric tangent of α.

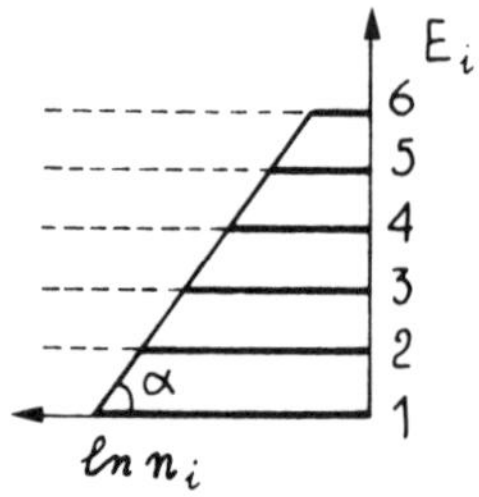

- Consider the two limiting examples

$T = 0$

Only one level is occupied, so there is maximum order.

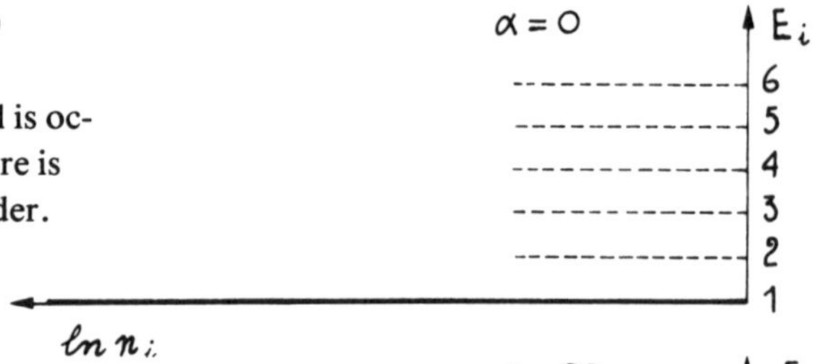

$T = \infty$

All levels are equally occupied, so there is maximum disorder.

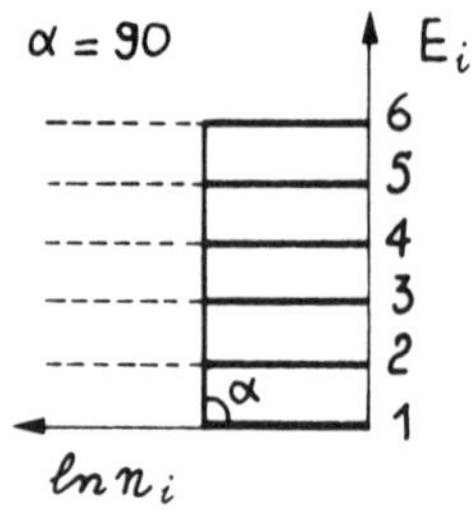

- Note that whereas at equilibrium T is an actual indication of disorder, Boltzmann's distribution law no longer applies outside equilibrium, when even if there is disorder, no temperature can be identified! Thus T is a characteristic parameter of the statistical distribution at equilibrium.

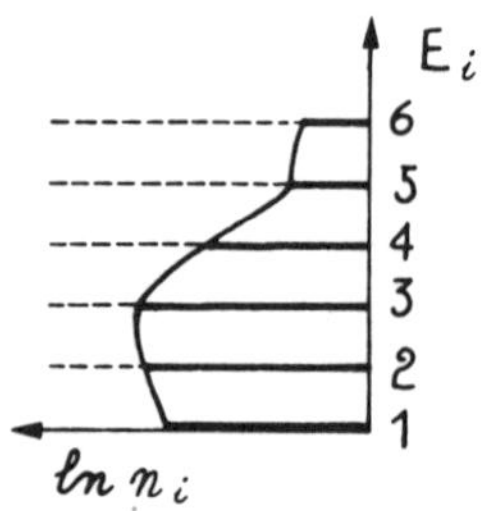

gle straight line whose slope is proportional to the temperature, if the Boltzmann distribution law holds. Let us try to verify this using two extreme examples. Indeed, at a temperature of absolute zero, the slope of the straight line must go to zero, and this means that only one level is occupied; thus the situation is one of maximum order. On the other hand, at infinite temperature, all levels have equal populations, because the trigonometric tangent of 90° is infinite; thus the situation is one of maximum disorder.

Although this reasoning enables us to identify the temperature at once, it likewise points to the limitations of this concept. Suppose we have a situation in which all levels are randomly populated. It is evident that in such a case we cannot draw a straight line that joins the ends of the segments, and so it is out of the question to speak of temperature. So, unless we are dealing with thermal equilibrium, when Boltzmann distribution does not apply it is impossible to identify the temperature. *Temperature is seen to be no more than a quantity characterizing the spread of the statistical population distribution among energy levels and only in conditions of thermodynamic equilibrium*. Thus temperature appears as a quantity that reveals disorder but certainly cannot measure its extent.

Statistical thermodynamics has taught us that only entropy can be used to measure disorder. To calculate entropy with our model, we should evaluate the population distribution on fixed levels and then use the appropriate mathematical expressions given by statistical thermodynamics. Unfortunately, we do not usually know how to make such a calculation, whereas in some specific cases it can now be performed with the help of computers. We can make a direct calculation only when the external magnetic field is zero, because then there is only one occupied level.

Let us try to use this model again, this time to form a concrete idea of the free energy F, which will also play an important role in this book. By definition, at constant temperature

$$dF = dE - T\,dS.$$

Only part of the internal energy change is free to be used in an isothermal process, because another part is inevitably bound to the system (the term $T\,dS$). To show this, suppose we vary the internal energy of our model system, keeping in mind that this variation splits into a work term and a heat term, according to the operations we have already described. It follows that, if the transformation is isothermal, the work performed must be followed by population rearrangement, so that in a logarithmic plot the ends of the segments that measure these populations lie along a straight line parallel to the first. For instance, if the magnetic field is increased, the population must decrease at the highest levels and increase at the lowest. Thus *not all the internal energy variation can be exchanged with the outside in the form of work,* because the thermal portion corresponding to the term $T\,dS$ remains bound to the system. The inevitable population rearrangement needed to keep the temperature constant includes just this exchange of heat. This concept will be illustrated in further detail in the next section, where we will examine the concrete example of adiabatic demagnetization.

1.4 Order in a Perfect Spin Gas

One good way to examine the differences between temperature and entropy, two quantities sensitive to disorder, is to proceed through a series of operations called adiabatic demagnetization. Suppose we isothermally increase the external magnetic field acting upon our model system so that the entropy must decrease because the spins become more ordered. At this point, we thermally isolate the system and then remove the magnetic field so there can be no further variations in entropy; we shall see that the temperature will drop simultaneously. The temperature and entropy variations in these two operations occur separately. Now let us take a detailed look at what happens to the spin populations, as shown in Box 1.E. In the first situation (A), the system has been exposed to a magnetic field smaller than the one to be applied next (not quite zero, because otherwise the group of segments would all merge into one segment, and this would cause some complications in graphing. If the magnetic field is increased, the energy levels rise because work has been done and the temperature goes up at once, as the figure shows. Since the system is not isolated, it must return to its original temperature below the constant high value of the applied field. This final situation is labeled B in the figure; note that in this kind of plot equal temperatures are shown as straight lines with the same slope.

Now let us thermally isolate the system (this will not cause any population variation) and then take the magnetic field back to its original value. In this last operation, the populations cannot change because no heat is exchanged, so it is possible to reach the final situation labeled C, where the spacing between levels is the same as the spacing in A, and the populations equal those in B. We can see that the temperature in situation C has dropped, and the system has cooled off after these two steps.

This model system offers a glance at the great difference between a transition at constant temperature and another at constant entropy, using an ingenious ordering process that works by applying an external field that is later removed. We explained in the previous section what it means to perform a transition at constant temperature: It means keeping the spread of statistical population distribution constant among energy levels during the

Box 1.E

ADIABATIC DEMAGNETIZATION OF THE SPIN GAS

- Consider the entropy S of a perfect spin gas in a field H.
- As H increases the entropy drops because the spins become ordered along H; let us draw the population distribution in the transition

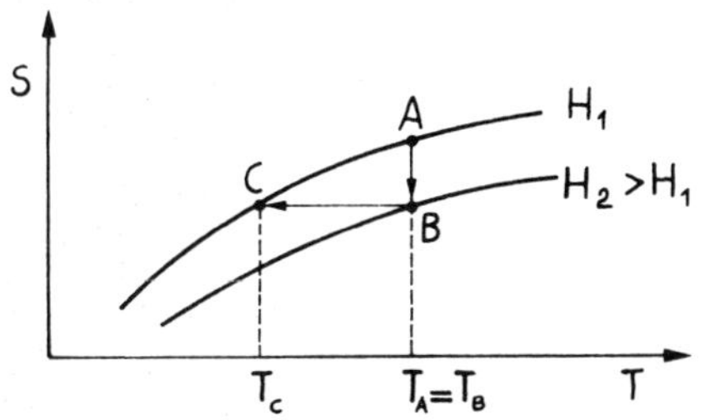

A → B → C

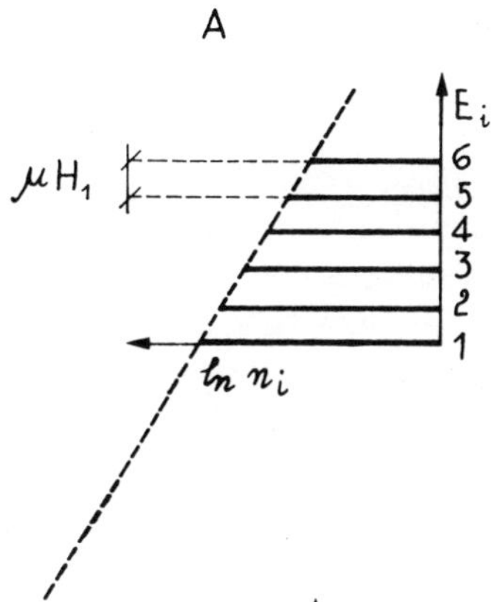

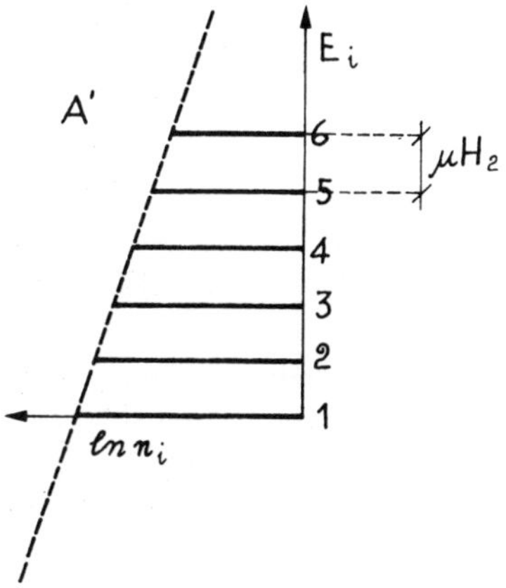

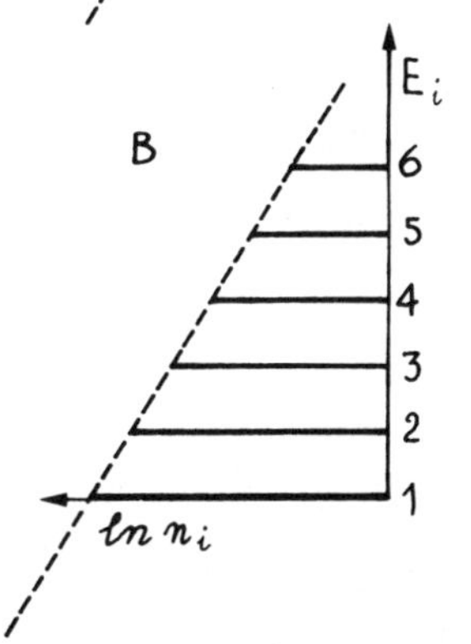

note that state A', obtained as soon as the field H_2 is applied, is outside equilibrium and thus cannot appear on the S, T-plot.

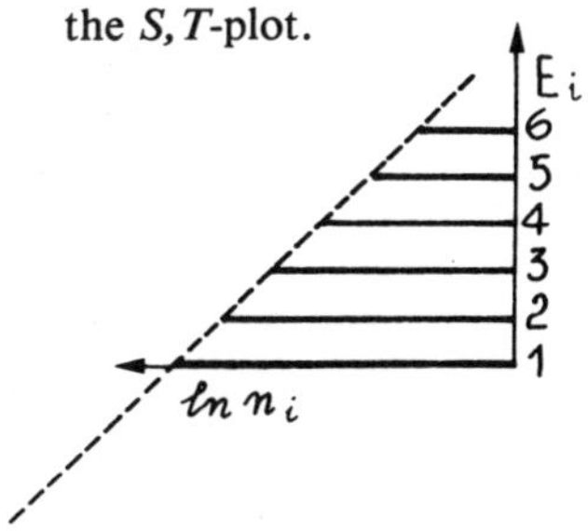

Box 1.F

ENTROPY VERSUS TEMPERATURE NEAR ABSOLUTE ZERO

- Compare the behavior of entropy S as a function of the temperature T and of the external parameter P in two different systems seen below.

- In a hypothetical system (a) that does not obey the third law of thermodynamics, order can be created by using a finite series of operations A → B → C such as those considered in Box 1.E.

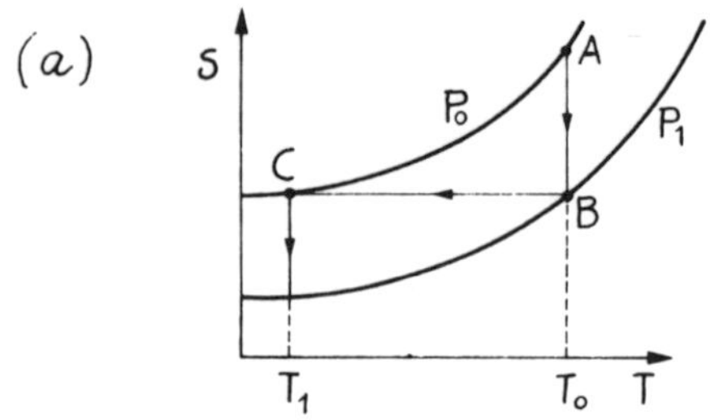

- In a real system (b), absolute zero cannot be reached using a finite series of such operations, because of the specific-heat behavior as a function of the temperature.

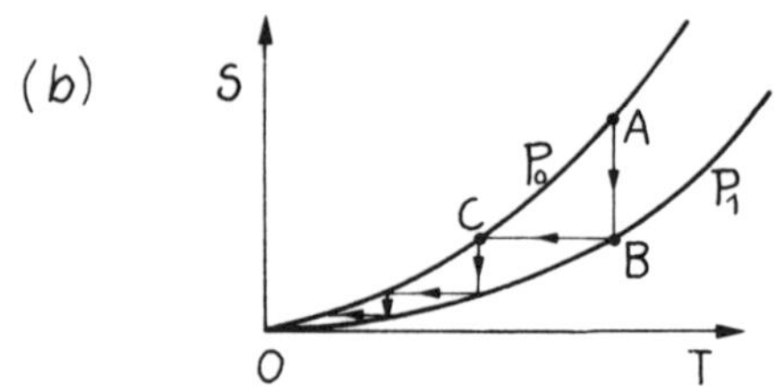

- If the specific heat could remain constant, which is impossible in a real system, the entropy at absolute zero should reach infinity, because

$$S_T - S_O = \int_0^T \frac{C}{T}\, dT = c\ (\ln T - \ln O) = \infty$$

transition. Now we can see that *in a transition at constant entropy the population of each level must be kept constant,* while the height of each level can vary. Only by keeping the population of every level constant can the disorder intrinsic in the distribution remain constant. Evidently, entropy depends on the number of levels but not on their energy value; in our plot, entropy is not linked to the height of these levels. This is an important conclusion.

Now we know what an adiabatic demagnetization is, and we can proceed to outline its remarkable experimental importance as a means to reach the lowest temperatures in a magnetic system. When working with paramagnetic ions sufficiently well separated by hydration water molecules, as in alum crystals, it is relatively easy to reach temperatures of one-hundredth degree absolute. With nuclear spins, and thus with magnetic moments that are weaker by at least 10^3, temperatures in the range of the nano degree have been reached for a short time. We are forced to use progressively weaker magnetic moments to reach progressively lower temperatures because we must keep the spin gas far from becoming *real*. By real, we mean that the interaction between neighboring magnetic moments becomes comparable to the energy of a single magnetic moment interacting with the external field. As we shall soon see, the entropy of a real spin gas is vastly different from the entropy of the perfect spin gas we have been considering here, because the former approaches zero far from absolute zero. This is consistent with the third law of thermodynamics, which is sometimes formulated by stating that it is impossible to reach absolute zero using a finite series of operations with a real system. Thus a series of ABC-type operations such as those we examined for adiabatic demagnetization can never allow us to reach absolute zero in a real system. This will become even more evident if we look at the two systems shown in Box 1.F.

1.5 Negative Temperatures

We can make another interesting comparison between the significance of temperature and that of entropy, by studying the behavior of a particular system of magnetic nuclear dipoles. The results of such a study lead us to extend the previously considered notion of temperature to include negative temperature as well. Indeed, it should be easy to accept such a generalization after having read the previous paragraphs. In the plots shown so far, the slope of the straight line that joins the population segments can also become negative, and this situation can also be described by saying that the system has a negative temperature. Of course there is the experimental problem of increasing population at the higher levels, but if the straight line expresses a true distribution at equilibrium, then such a generalization is perfectly legitimate.

Negative temperatures were first achieved in 1952 at Harvard, when an experiment was carried out on a lithium fluoride crystal, where the nuclear spin system takes quite a while to reach equilibrium with its thermal bath. Because of this property, there is enough time to experiment with an isolated spin system, which behaves like the perfect gas model of magnetic dipoles. The lithium nucleus has a 3/2 spin factor, and thus the spin system populations arrange themselves on four different levels (see Box 1.G). If the

Box 1.G

NEGATIVE TEMPERATURES OF THE SPIN GAS

- Let us generalize the diagrams of the previous boxes for this situation of thermodynamic equilibrium:

  $\alpha < 90°$

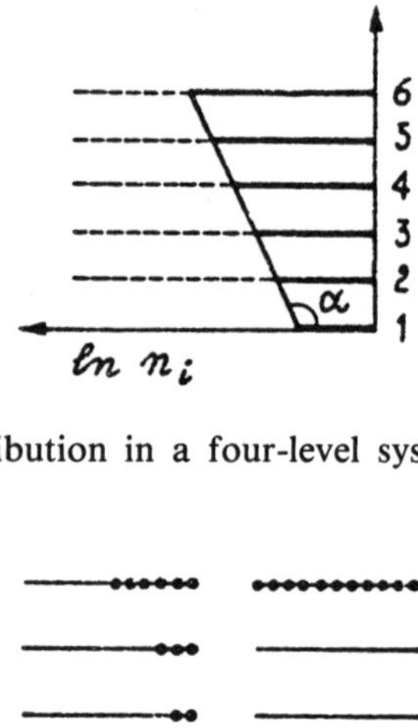

- Schematic plot of the population distribution in a four-level system such as lithium fluoride:

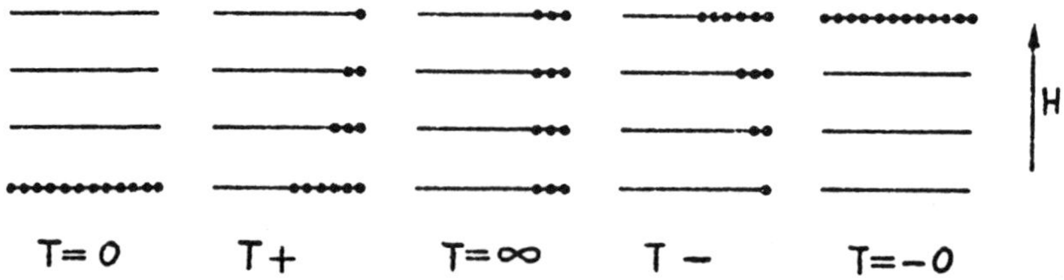

- Note that order is greatest when only one level is occupied, because only then is entropy equal to zero, and that this occurs in the two situations

$$T = +0 \qquad \text{minimum energy,}$$

$$T = -0 \qquad \text{maximum energy.}$$

external magnetic field is rapidly reversed, the magnetic system that should previously have populated mainly its lower levels ends up in the more populated upper levels, because the magnetic moments now find themselves arranged in the opposite direction of the field. On the basis of suitable experimental measurements, it was next shown that the whole population distribution evolved gradually, first going through a state in which all the populations were equal, and then settling into a state corresponding to the final equilibrium of the system in contact with the bath. The temporal evolution of this phenomenon can be described by saying that there are two systems in the lithium fluoride crystal—one being the lithium nuclear spin system and the other being the lattice system of the ions that give rise to the thermal bath—and these two systems have different temperatures. After the nuclear spin system has been reversed, it must return to the situation imposed upon it by the lattice, because of the latter's greater thermal capacity. The interaction between the two systems is slow enough to allow for measurement of the populations on single levels, but the microscopic details of this interaction are omitted here.

Now let us see how temperature and entropy behave differently during the experiment just described. A look at Box 1.G shows that *entropy must be zero when only one level is full,* and this occurs both *at the temperature of normal absolute zero* (the situation we discussed in the last section) and *at the lowest negative temperature.* Indeed, the system is at its "hottest" in this latter situation, because it can shed heat more readily when it is richest in energy. We can also see that negative temperatures are not the ones found below normal absolute zero, as they would be for the usual coordinates, but are those above infinite temperature. We should now have a clear understanding of the meaning of temperature as the parameter that characterizes only statistical distribution among the energy levels. Moreover, we should likewise have reevaluated the third law of thermodynamics.

Before moving on, we should point out that few real systems can present negative temperatures. The experiment we have just seen is possible only when the number of energy levels is finite, as happens in the case of a quantized spin system. This no longer holds true in the case of a molecular gas, a liquid, or a solid, because the number of translational, rotational, and vibrational energy levels is infinite. Since it is necessary to cross the infinite temperature on the way from a positive to a negative temperature, this means that infinite energy must be supplied to the system to populate all its infinite energy levels equally. Obviously, infinite energy is never available.

1.6 Spontaneous Order in a Real Spin Gas

At this point, we shall have to leave our model of the perfect spin system—so useful for understanding the statistical thermodynamic meaning of several fundamental quantities—and move on to the corresponding real system that includes the interaction among the single spins of the system. If we introduce magnetic forces among the magnetic dipoles (in our previous model system we deliberately overlooked such forces), we must obviously move the system's energy levels and consequently change the relative statistical populations with respect to the value they would have had at the same temperature in the perfect gas model.

Once the energy levels of the system have been identified, statistical mechanics gives us the possibility of calculating thermodynamic functions by determining the so-called partition function; this is the sum of the statistical weights of each level, extended to all levels of the system. In practice, this is a highly complex calculation for a real system, given the enormous number of terms to be considered and the impossibility of working with simple analytical functions. In the case of a real spin system, the problem can be easily solved in one dimension by postulating an interaction other than zero for the first neighbors alone. This approximation is called the Ising model. In two dimensions, the problem becomes extremely difficult, but it was solved in 1944 by Onsager, whereas it has yet to be solved in three dimensions.

Given these difficulties, we shall have to adopt an intuitive and qualitative method here to establish the existence of a rather general phenomenon, *the spontaneous ordering of a real system below a critical temperature.* Going back to our previous discussion of adiabatic demagnetization, let us now include the Weiss field, a mean internal field caused by the magnetic interactions among dipoles. It is easy to understand that this internal magnetic field can operate as the external magnetic field did, to lower the entropy of the system. As we shall soon see in detail, it is essential that this occur only below a particular temperature, which is the critical one called the Curie temperature for magnetic systems.

Therefore, let us think of a dipole in the real system, namely in the field of all its neighbors. It is obvious that, if the result of this local internal field is not zero, the dipole we are considering will line up along this resulting field to lower the system's energy. This alignment is usually opposed by thermal agitation; this is the same event that occurred when the external field was in operation, as we saw when discussing the Langevin curve. As before, order and disorder are in competition and they meet in terms of energies. Thus, if we operate in the absence of external fields and if the mean internal field is the only force responsible for establishing order, at the Curie temperature the thermal energy must be equal to the energy of the magnetic dipole in the internal field, as shown in Box 1.H. Then by measuring the Curie temperature and knowing the magnetic moment of the dipole, we can evaluate the internal magnetic field. In the case of paramagnetic salts, this turns out to be the correct order of magnitude.

Box 1.H

SPONTANEOUS ORDER IN A REAL SPIN GAS

- If the magnetic moments of the spins interact, the internal field H^* that originates will act like an external field to order the spins by lowering the entropy in the system.

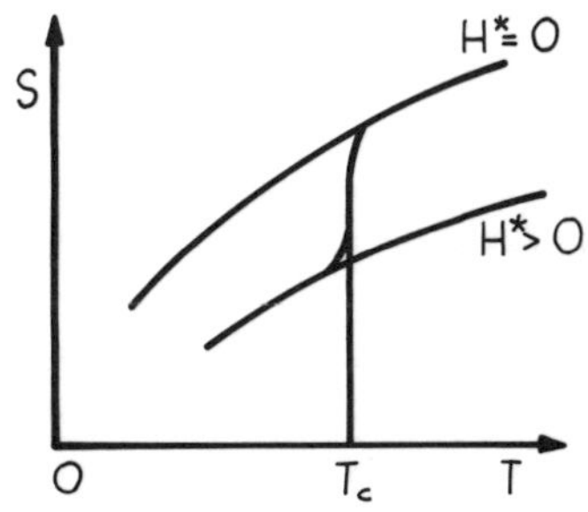

- This must occur below a certain temperature T_C called the Curie temperature when the randomly disordered thermal energy is almost the same as the internal field energy that aligns the dipoles

$$kT_C \simeq \mu H^*$$

- The spontaneous magnetization M that exists only below T_C shows the typical trend of all cooperative processes with a saturation value M_s

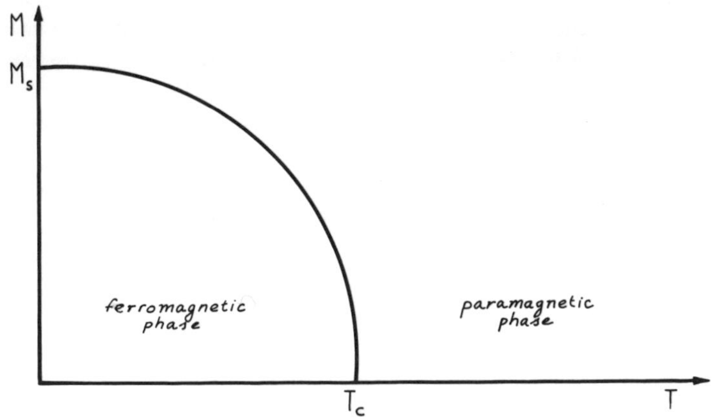

Now look at the fascinating process of *spontaneous magnetization,* presented by the dipole system below the Curie temperature, and see how it can be easily understood by extrapolating the process of adiabatic demagnetization to the real system. This must occur, because the order that the external field formerly produced in the ideal system is now provided by the system itself through its own mean field. Of course, this process involves cooperation in the sense that, if a fluctuation creates some nonzero component in one direction, all other dipoles will line up in the same direction to reinforce it. In other words, near the Curie temperature, if the system is further cooled off even slightly, a catastrophic process is spontaneously set off, arranging all the dipoles of the system in one random direction. Just as it was necessary before to remove heat in the perfect spin gas, to lower entropy at a constant temperature and raise the populations at the lower levels, now an exothermal reaction is developed to enable the system to get rid of its heat content at the Curie temperature. This effect, called magnetocaloric, occurs because an internal field in isothermal conditions leads to the rise of all energy levels and therefore to the increase of populations at the lower levels.

Naturally, the magnetization of the real system is always the sum of all magnetic moments in the system; this is a macroscopic and measurable quantity that gives the best indication of the amount of order in the magnetic system. Indeed, although magnetization is zero above the Curie temperature, because the mean value of the resulting field of randomly arranged dipoles is zero, at the Curie temperature magnetization suddenly greatly increases and quickly becomes asymptotic to its value at absolute zero. This saturation value is obviously reached when all dipoles are parallel to each other, and is the perfect order we must always expect to find at absolute zero. The presence of the internal field changes that picture; now this order starts at a much higher temperature, and for this reason, we are correct to call it the spontaneous ordering of matter.

What we have just seen is called a *ferromagnetic transition;* the two phases separated by the Curie temperature are called, respectively, the ferromagnetic phase (the ordered one found at the lower temperature), and the paramagnetic phase above. Similarly, we call a "ferromagnetic domain" a region of matter that is completely ordered in one direction, a region that does not usually coincide with the size of a monocrystal. Actually, even if the kind of ordering we have just described really does take place in iron and other elements of the transition group, in these cases the interaction among electronic spins is quite a bit more complex than that taking place among elementary magnetic dipoles. On the basis of the value of an isolated magnetic moment and the space distribution of the moments of metal ions, the intensity of the local field in a ferromagnetic domain should turn out to have a much lower magnitude than the one that has been measured. This is comprehensible only in terms of the band theory of the transition metals, a topic we shall not consider here. Nevertheless, the term "ferromagnetism" is still used to label an ordering process caused by mere interaction among spins, such as the one we considered in this section. This process does occur in nature in paramagnetic salts at the lowest temperatures.

1.7 Spin Waves as a Quasi-Particle Gas

The system of interacting spins seen in the last section can be used again to introduce another important notion, the spin wave. This notion can in turn be generalized into the notion of the *quasi-particle, which is the element of disorder in a material system that is already very ordered.* This concept is now of great importance in describing a wide variety of situations both in and out of equilibrium.

The best way to approach this subject is to point out that the ground state of a group of classical interacting magnetic dipoles is reached when all dipoles are arranged in the same direction. For simplicity, we will consider a system where only two positions are allowed. In this instance, the first level of higher energy corresponds to a situation in which only one dipole is pointing in the opposite direction, because work must be done to turn it against its neighbors' field. As we know, one of the general results of quantum mechanics is that the lowest energy at equilibrium is reached by reducing the space gradient of the energy itself—i.e., distributing it over a wider space. So we can lower the energy of a quantum system if we distribute the directional change of a dipole over a number of dipoles in a wide enough space (see Box 1.I). The laws of quantum mechanics also state that the extra energy of this "disturbance" over the ground state must be quantized. This notion, exactly analogous to the light photon concept in radiation theory, shows that this quantum of energy can be labeled a "quasi-particle" called a magnon. According to Planck's law, the energy of this quasi-particle depends on the corresponding wavelength; therefore, the lower its frequency, the lower its energy and the higher its wavelength will be.

The quasi-particle concept is useful because the system becomes remarkably simpler to describe when it is very ordered. In this case, we can imagine that the system is no more than an empty container with the dimensions of the body, in which some material points (the quasi-particles) move about as though they were the molecules of a rarefied gas. In other words, *the disorder is restricted* to a few places because there is not much of it. If we remember the wave–corpuscle duality of quantum mechanics, we see that this model will be useful only as long as the waves do not overlap and become confused with one another.

There is a great deal of experimental evidence of the existence of quasi-particles. In other words, it has been seen that some disorder can be restrict-

Box 1.I

THE EARLY DISORDER DESCRIBED AS SPIN WAVES

- Let us introduce an element of disorder into a classical system of interacting spins:

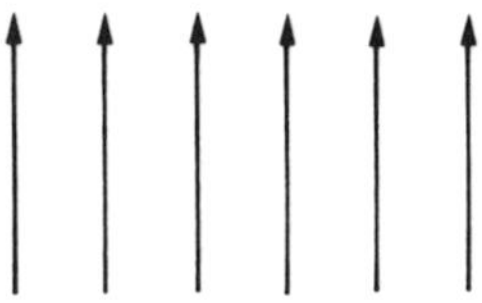

Ground ordered state

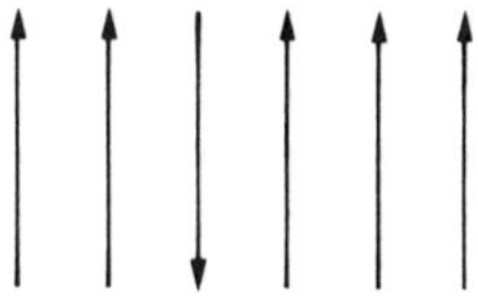

Excited disordered state

- A general result of quantum mechanics states that energy drops when it is more widely spread, thus reducing its gradient. So let us form this wave distribution of low-energy $\mathcal{E}$:

$\varepsilon = h\nu$
$\nu =$ frequency
$\simeq 1/\lambda$

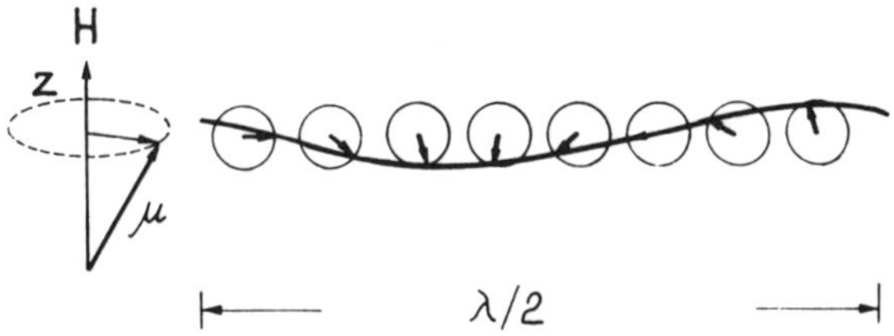

Stationary wave seen on an *x,y*-plane

- To generalize, we can draw disorder in a highly schematic way:

Paramagnet

Ferromagnet

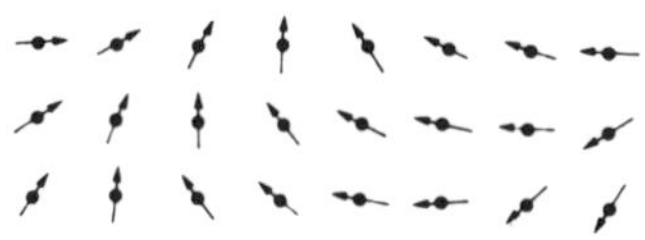

ed to one point and that this point moves according to the laws of mechanics. For instance, we can create a magnon by letting a neutron beam go through a ferromagnetic substance at a lower temperature, so that some of the neutrons interact with the ordered magnetic dipoles and give them energy. In this process, the neutron is detoured, as though it were a billiard ball striking an immobile ball and setting it into motion. The collision laws of mechanics make it possible to establish the amount of energy taken away by the magnon thus created. Just as a photon gives up energy to a metal electron in the photoelectric effect, so a neutron gives up energy to a magnon. This implies that energy has been localized in one spot in the electromagnetic field in the case of the photon, just as the spin-wave energy has been localized in the case of the magnon.

For the purposes of this discussion, it will be useful to picture a quasi-particle as the *correlation between thermal fluctuations*. If the temperature is not zero, some magnons must be present, because there is disorder and they are the carriers of disorder; to be more exact, the average value of magnon frequency corresponds to kT/h. If there were no magnons, this directional disorder would be randomly distributed in the form of thermal fluctuations over all dipoles in the system. The existence of the magnon as a collective wavelike aspect in the motion of these dipoles implies a space–time correlation between these thermal fluctuations.

To play with words a bit, we could say that the first disorder created in an almost ordered system is itself "ordered," because a correlation implies some sort of order. Naturally, this is not a permanent temporal correlation between localized dipoles, and thus it is quite unlike the correlation displayed by a stationary wave, which expresses authentic coherence. Quasi-particles display only a statistical kind of correlation; we shall return to this important concept.

1.8 Molecular Aggregate Models

At this point an analogy comes to mind, linking our observations on the spontaneous ordering of a magnetic dipole system and what is already known about the various states of aggregation. Indeed, this is no superficial analogy, because the perfect gas of magnetic dipoles is to the ferromagnetic domain what the perfect gas of molecules is to the condensed molecular state. In the former case, the forces operating between neighbors are magnetic, whereas in the latter the forces are electrical (either in the form of forces between permanent electric dipoles or of London forces between polarizable electrons). Of course, the same features are displayed by the molecular aggregates, except that the magnetic field is replaced by pressure. In this case, we can introduce an internal pressure caused by molecular interactions, which is analogous to the internal magnetic field; and we can identify condensation, which occurs when the temperature of a vapor is lowered below the critical temperature, with the process of spontaneous ordering that now leads to the denser stage.

To take this analogy one step further we should first identify the energy levels of the molecular system. Unfortunately, this is easy only for a low-density vapor or a solid at fairly low temperatures. The reason for this is that, in the case of a vapor, the only energy levels are the transitional levels of a nearly free particle in a box and, because of the box's macroscopic dimensions, these energy levels are so close to one another that they become equally populated. This is perfectly analogous to the equal population of the degenerate magnetic levels in the absence of an external or internal field. In the case of a solid at low temperatures, we might approximate atoms to quantized oscillators, so that the levels of the system are the equally spaced ones of a harmonic oscillator in a suitable force field. But for a liquid, it is hard to obtain a simple picture of how the energy levels might be arranged. We could compare the liquid to a solid in which the atoms are subjected to a field that varies from place to place, to account for the greater local disorder of the liquid—so in place of one single level of the solid we should have a continuous band for the liquid. Apart from the scant utility of this liquid model, it is clear that it is simply impossible to construct one model to represent all the different random situations that might occur in a liquid, ranging from compressed gas to something resembling a crystal, close to the point of fusion.

Today, an oft-used representation of the way the molecules of an aggregate distribute themselves in space is the space correlation function, specifically the radial distribution function shown in Box 1.J. As the vector radius increases from a reference molecule, this function tells us the probability of simultaneously finding a second molecule at the distance indicated by the vector radius, assuming that completely random distribution is equal to one. It can be seen that this probability is zero in the region where the presence of the second molecule is ruled out by mere geometrical reasons of overlap; otherwise, it is a variable function, with maximum and minimum points. In the case of a liquid, these maxima are found at the nodes of the corresponding crystal lattice, and the entire curve can be determined by x-ray or neutron diffraction. This example shows us again that *the order of a molecular aggregate can be correctly expressed as the correlation among its own parts,* and more precisely here, as a permanent correlation among the microscopic fluctuations of local density that would otherwise be completely random. Therefore, it can be said that there is a strong correlation among the small-amplitude fluctuations of atomic positions in a crystal at low temperatures. This correlation causes the formation of phonons, whereas in a rarefied gas there is only a slight correlation among the molecule positions. The system's internal forces are a function of distance, and

Box 1.J

SPACE-CORRELATION FUNCTIONS IN FLUIDS

- The space-correlation function $G(R)$ expresses the probability of finding another molecule at a distance R from a reference molecule. The figure shows its construction for the case of hard spherical molecules.

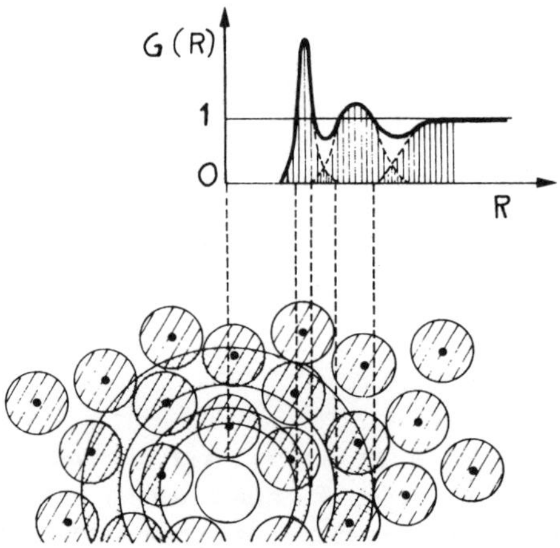

- The correlation function $G(R)$ depends upon the potential energy $U(R)$ of the intermolecular field. The figure below shows the behavior of $G(R)$ for two different potentials, one for a model A of hard impenetrable spheres and another one B which is more realistic because it includes a long-range attractive component. (See Box 2.A.)

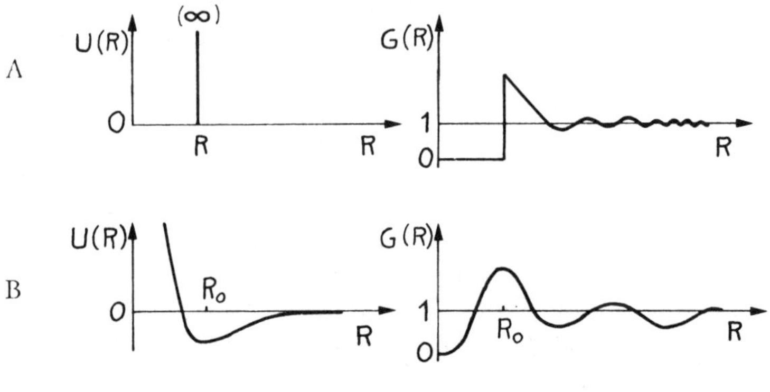

Box 1.K

ANALOGIES BETWEEN MAGNETIC AND MOLECULAR SYSTEMS

Magnetic systems	*Molecular systems*
Magnetic dipole	Atom or molecule
External field	External pressure
Magnetic levels	Molecular levels
Weiss internal field	van der Waals field
the Curie temperature	Critical temperature
Spin waves in domains	Lattice phonons
Spin inversion	Point defect

they produce this correlation to varying degrees. The best way to represent the intermediate state between order and disorder in a liquid is probably to determine the correlation between these space–time atomic fluctuations. The radial distribution functions mentioned above are the two body elements into which the correlation in a many-body system can best be split. These radial distribution functions can be reconstructed on the computer with dynamic-simulation techniques by postulating different types of intermolecular forces. This gives a clear demonstration of the importance of force-field details on these statistically significant quantities that bridge the gap between the microscopic and macroscopic worlds.

To go back to molecular aggregates, it remains evident that *the analogy between the Weiss field and the internal field of intermolecular forces,* called the van der Waals field, *is truly a good one.* Box 1.K shows that this analogy can be extended to wavelike distributed disorder, which in the molecular system brings us to the concept of phonons in correspondence with the concept of magnons; in addition, the analogy applies to localized disorder, such as in the case of a crystalline defect in the molecular system, which corresponds to one-spin reversal in the magnetic system. Since we have already sufficiently discussed the wide-scale distributed disorder, let us now take a brief look at some states of localized disorder in solids. When we speak of localized crystalline disorder we are usually referring to situations that can be found even in monocrystals grown with the greatest care, when the atoms sit in an interstitial position or when they leave some lattice nodal

Box 1.L

LATTICE DISORDER IN CRYSTALS

- Two-dimensional drawing of a simple crystalline solid with some typical point defects.

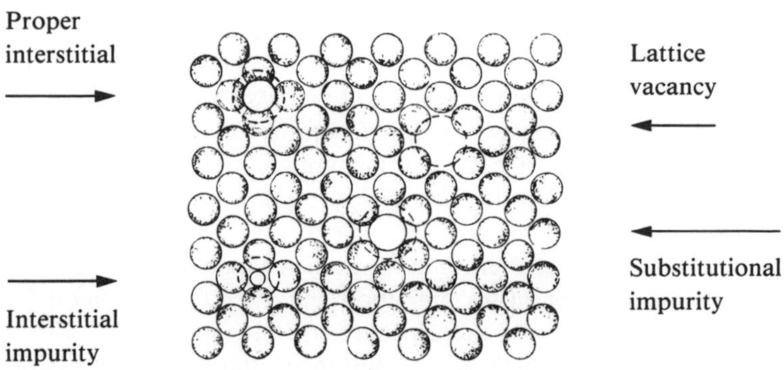

- Compare the above localized situations with the distributed thermal disorder of a crystal at high temperature:

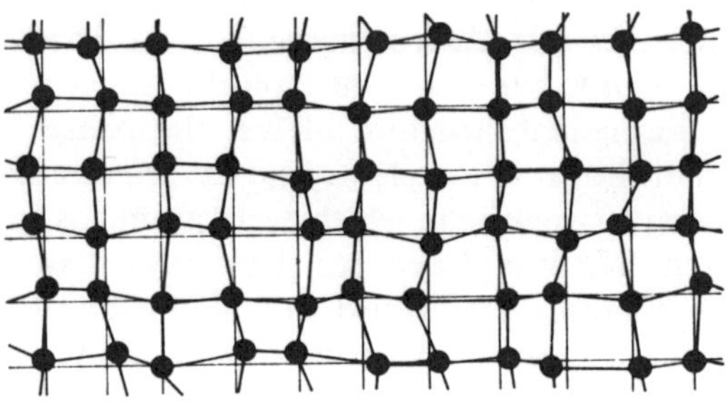

positions empty. It is clear that these localized defects must display higher energy levels than those corresponding to the perfect-crystal ground level, because some work must be done to remove an atom from its position of stable equilibrium in the lattice. It is also clear that these potential energy levels essentially depend upon the geometry of the lattice and upon the local interactions among atoms, as seen in Box 1.L.

We should expect the number of localized defects to increase as the temperature rises, because the Boltzmann law states that in this case the population of a higher level must increase. The case of a defect in a crystal can often be described using a two-level model system, which is completely analogous to a 1/2 spin gas in a constant external magnetic field. With the Langevin curve in mind, we can conclude that there must be a well-defined number of lattice defects in the crystal at every temperature, and that therefore the crystal will only be truly perfect at absolute zero. Note also that the number of defects is also low when the temperature is low, because of the high energy value that appears at the exponential of the Boltzmann factor, and thus every defect can be considered as an independent entity; in this case the previous model rings true. On the other hand, at higher temperatures we must be prepared for the number of defects to increase until there is a good chance that they will be close to one another. At this point, disorder can rise catastrophically because of the cooperative effect of the interactions; for instance, it becomes easier and easier to form vacancies next to one another in the lattice, until the whole crystalline structure collapses.

In molecular aggregates there may be another type of disorder, one that depends not upon temperature but upon the disordered structure of the aggregate, in an intrinsic and therefore irreducible manner. Of course, this does not contradict the third law of thermodynamics, because when the temperature is lowered, this disordered structure becomes unstable and thus transforms into another that does not have this intrinsic disorder. Nevertheless, if the temperature is lowered at a sufficiently rapid rate, these defects in the now metastable structure can become frozen and the amount of disorder they introduce is measured by *a residual entropy that does not disappear at absolute zero*. One typical example of this behavior is displayed by the hexagonal structure of ice, the phase found at atmospheric pressure and shown in Box 1.M. The oxygen atoms are larger than the protons, forming a structure in which every atom has four neighbors in tetrahedral position. The bonds between these oxygen atoms are filled by protons, but every proton must be next to only one oxygen atom in order to form the configuration of the water molecule locally. It is obvious that this condition can be reached in several ways within the same lattice structure. The presence of these defects causes an irreducible disorder that is measured by the residual entropy of ice at a low temperature, in perfect agreement with the theoretical value of entropy calculated as the logarithm of the number of ways to arrange all the protons near all the oxygen atoms. Naturally, at the lowest temperatures, this phase of ice becomes unstable and transforms into another phase that no longer displays this kind of structural disorder.

Box 1.M

STATISTICAL DISORDER IN ICE

- Ice structure I (hexagonal crystalline phase stable under normal conditions)

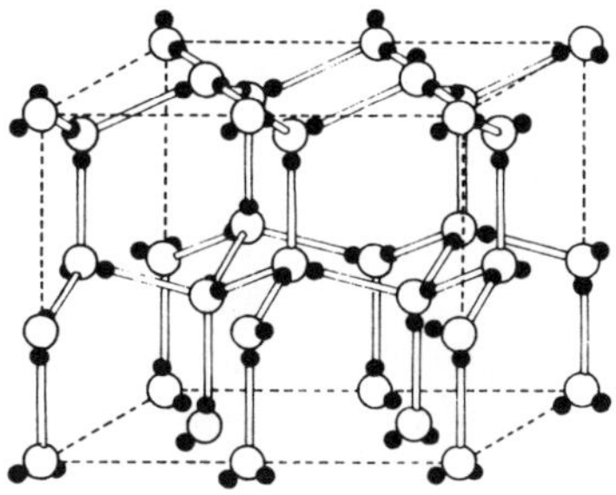

- Note the presence of one proton (black dot) on every bond between two oxygen atoms (white ball).

- Note also that in order to complete a water molecule, each oxygen must have two protons close to it.

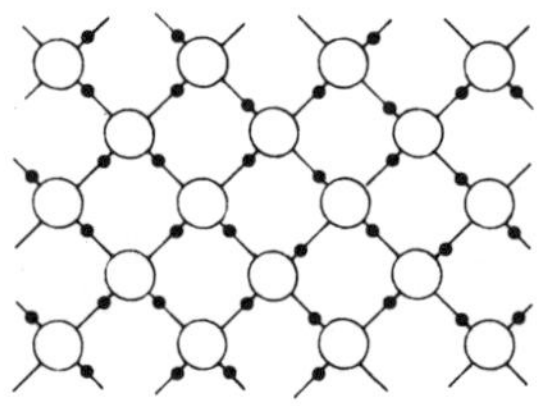

- Consequently, there is an implicit disorder in this structure, as drawn here in two dimensions.

1.9 The Einstein Model

The historical origins of the Einstein model are of great importance. Contemporary classical theory showed that, when a solid was approximated as an oscillator system, the specific heat should equal 3 R, where R is the constant of the perfect gas equation, for all monatomic substances; whereas experience showed somewhat lower values as the temperature dropped. In 1907, Einstein proposed that the vibrational thermal energy at low temperatures might be unequally distributed among the system's oscillators, inasmuch as energy was no longer sufficient to populate all the oscillators with at least one quantum. The calculation of specific heat is immediate if we

Box 1.N

THE EINSTEIN AND DEBYE MODELS

- In Einstein's model, every atom is located in its neighbors' mean field, and it vibrates around the center of its own three-dimensional cell.

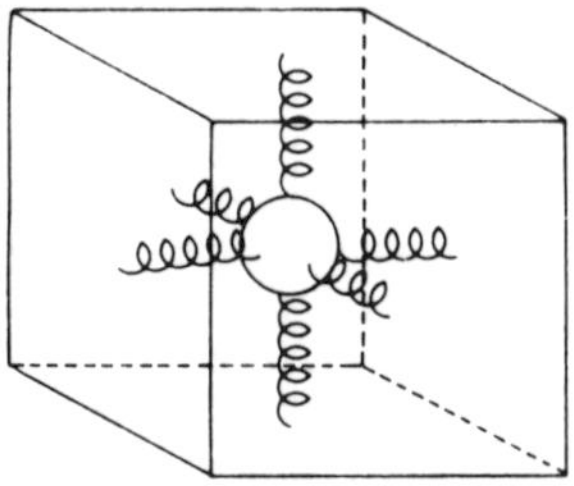

- Compare the above Einstein model with the Debye model for a group of four atoms in two dimensions (drawing the atoms as black dots and the forces as springs).

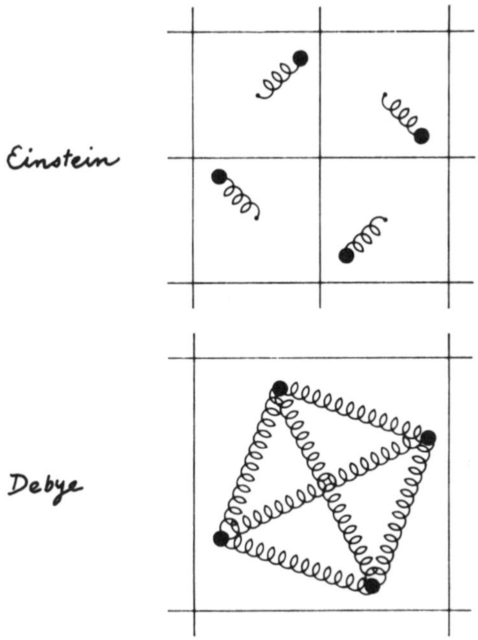

accept Planck's distribution law for oscillator energy. It appears that the quantum correction is important only below a characteristic temperature T_E, called the Einstein temperature, which is found by postulating the average thermal energy of the oscillator to be equal to the energy of the vibrational quantum. This evidently means that below this temperature not all the oscillators can be populated by one quantum. The value of T_E is experimentally deduced from the temperature below which Dulong and Petit's law no longer holds; this value turns out to be close to room temperature for most solids. From this value of T_E we find the lattice frequency falls in the infrared region. The corresponding wavelength is about one angstrom long, and this means that every atom vibrates essentially within one lattice cell (see Box 1.N).

At this point, let us see what this Einstein model implies. Every atom vibrates within the average potential cage caused by the presence of nearby atoms; when this potential is approximated to a parabola, it generates the harmonic motion of the oscillator. The harmonic oscillator energy is then quantized on equidistant energy levels, and the first of these is the ground state, where the energy is zero. This scheme was later modified by quantum mechanics, but that does not alter the calculation of specific heat because the zero-point quantum energy cannot be exchanged. In this model, order and disorder are defined as the number of oscillators in the ground state and in all of the higher vibrational levels, respectively. Consequently, at absolute zero all the oscillators are in the ground state, and as the temperature rises this number can only decrease. But the most fragile part of the Einstein approximation is the assumption that every atom oscillates in its own cell, independently of the other atoms' actual positions in their nearby cells, because the correlations between positions of nearby atoms have been deliberately overlooked. *This incoherence in atomic motion,* and thus in the kinetic energy, *is precisely what characterizes thermal energy.* So the model achieves Einstein's goal, to explain the behavior of specific heat, but it completely fails to account for other phenomena and particularly to describe sound. Indeed, a sound wave is the correlated oscillation of a great number of atoms around their equilibrium position, but the atoms in the Einstein model each vibrate in their own cells with no phase correlation between even the nearest of them.

Before going on, we should mention another aspect of the Einstein model that has contributed to important developments. Note that in our discussions we have never specified which atom of the lattice possesses a vibrational quantum and which was in the ground state. Implicitly, we assumed that *the vibrational quanta are not localized,* each on its own atom, but that they are distributed throughout all the atoms of the solid as a collective kind of energy. This can also be explained by saying that a vibrational quantum jumps from one cell to another at a speed close to sound, a value with which atomic motions in a dense phase usually propagate. So the vibration of the Einstein oscillators that we saw contained in one lattice cell should not be described as a stationary wave but as a traveling wave of the same frequency, propagating with the speed of sound. Thus we come close to the phonon concept, which we shall discuss further in the next section, starting from a different model of a solid.

1.10 The Phonon Gas

The behavior of specific heat predicted by the Einstein model differed slightly from the one observed experimentally at the lowest temperatures. To improve the theory, Debye devised another model in 1912 in which quantization was no longer applied to the motion of a single atom but to the elastic wave of the microscopic body. An elastic wave expresses the *correlation in the vibrational motion of a large group of atoms,* and thus in this scheme the motion of one particular atom results from the combined effects produced on it by all the elastic waves and therefore by all the interactions in which the particular atom is involved. In other words, the complex nature of local atomic motions is replaced by the ensemble of waves that move that atom. Therefore, specific heat can be calculated as soon as the spectrum of elastic oscillations is identified, and this spectrum in Debye's model is approximated by a continuous spectrum conveniently cut off at high frequencies.

Now let's see why Debye's model should be better than Einstein's at very low temperatures. When the internal energy is low, Debye's model states that the waves of greatest statistical weight are the low-frequency ones, i.e., those with the largest wavelength. These waves can distribute energy throughout a wide space, unlike those of the Einstein model, which concentrate energy in one small spot like the lattice cell. Some general criteria of quantum mechanics, mentioned above with reference to spin waves, show that Debye's model ought to be better. Indeed, the most correct way to picture the disorder in a solid at low temperatures is by means of a group of waves of different frequencies, contained in an empty box with the same dimensions as the solid. This way, disorder is expressed by the very waves that move matter, and it naturally disappears toward absolute zero, where every motion must stop, according to the semiclassical point of view. Incidentally, we mention again that, according to quantum mechanics, this is not true any more because of the residual zero-point energy, which does not, however, contribute to thermal energy and entropy. Of course, a single elastic wave represents a coherent dynamic situation and thus expresses order, whereas a group of waves that spread out in every direction with different phases and wavelengths must depict disorder throughout the solid.

The waves we considered above belong to the frequency spectrum of modes of a continuous medium and are thus stationary waves. At this point, we should remember that a stationary wave can be decomposed into two waves of the same frequency traveling in opposite directions. For reasons we have already mentioned, we can *associate a quasi-particle to every traveling elastic wave;* this particle is the phonon, which is to the elastic wave what the magnon is to the spin wave.

We can offer a microscopic picture of the phonon in terms of the coordinated displacements of a group of atoms around their equilibrium position. Since a solid can present both longitudinal and transversal displacements along the direction of wave propagation, we must think of two corresponding types of phonons. Moreover, if the lattice is ionic, as for sodium chloride, the motion of opposing ions must give rise to an electrical dipole moment, the amplitude of which oscillates at the same frequency as the phonon. Furthermore, since an oscillating electrical dipole generates electromagnetic waves of the same frequency, we must expect an ionic lattice to radiate in the infrared region. In this way the thermal energy present in the solid medium as a phonon gas is dispersed throughout the surroundings as a photon gas, and the same holds true for the opposite process of infrared light absorption. Note also that *the phonon gas is an incoherent system* and it is disordered, so any process that effectively leads to a change in its density may be identified with heat. The phonon concept is particularly useful at low temperatures, when only a very small number of phonons are needed to describe disorder. In these conditions, the hotter spot in the solid is completely equivalent to the region of greatest pressure in a gas, and in both cases thermodynamic equilibrium is obtained when a uniform space distribution is reached. Finally, just as the flow of a rarefied gas through a pipe is mainly limited by the collisions of the gas molecules against the container walls, so the flow of heat at low temperatures is limited by the collisions of the phonons against the body's own surfaces.

Thus far, the phonon seems analogous to the photon, the quasi-particle that results from the quantization of the vibrational modes of electromagnetic field in a cavity. However, we must consider one important limit to this analogy. The photon gas is always a perfect gas, because the density of generally available electromagnetic radiation is so low that the quasi-particle photons are usually quite far away from one another. This can be expressed as the full validity of the superposition principle of electromagnetic waves, caused by the linearity of Maxwell's equations. But in the case of the phonon gas this is true only for low-energy elastic displacements as in the case of ordinary sound waves, namely those with a long wavelength. At higher temperatures, elastic displacements do not show a linear dependence upon the force, and the presence of anharmonic terms invalidates the superposition principle. Consequently, sound displays a wavelike propagation in a solid precisely because the superposition principle holds true at low elastic energies, but this cannot happen with heat because the elastic displacements have much greater energy. Heat can propagate as waves only in a region of a few lattice cells, whereas sound propagates throughout the whole body. *The impossibility of forming coherent thermal phenomena on a macroscopic scale* in a solid is perfectly in line with our idea of heat. To be more exact, we should say that phonons have a relative collision cross section that increases rapidly with frequency and thus high-energy phonons cannot propagate to great distances.

1.11 The Fermi Gas

One model that is important because it can be applied to the electrons in a solid is the perfect gas made up of particles that are indistinguishable from each other and obey the Pauli principle. This principle states that all the particles in the system must have different quantum numbers and that consequently no more than two electrons can be on each translational kinetic-energy level. This restriction in population density means that, even at absolute zero, the electrons cannot merge to one level, as happens in a quantized gas that obeys Boltzmann statistics (such as the spin gas considered above), or Bose statistics. Indeed, the electrons must now arrange themselves in pairs on all energy levels—from the lowest to the highest, which is the Fermi level at energy E_F. Naturally, *the entropy of this Fermi gas is zero at absolute zero, even if the system is made up of several occupied levels,* because we are dealing with a single state that can undergo no further ordering process. This is all in perfect agreement with the third law of thermodynamics.

Above absolute zero, the Fermi system's behavior might at first seem unexpected. Given the high E_F value for the more interesting cases, the possible energy exchanges with the outside barely suffice to move the electrons found close to the E_F threshold, because they are the only ones that can move to the higher empty levels nearby. The electrons located in the sea of levels below E_F must remain where they are; not enough energy can be given to them to take them to the empty levels above E_F. Thus at finite temperatures the electron system is always very ordered, precisely because of this "freeze" at the lower levels of Fermi's sea. Only the electrons near the E_F threshold will behave like an ordinary gas from the viewpoint of disorder, but the thermodynamic functions of the complete system of fermions turns out to be very different from those of the perfect ordinary gas.

Fermi and Dirac proposed the statistical thermodynamics of this model system in 1926. The model was immediately applied to the case of electrons in metals to explain the behavior of the thermal and magnetic properties that should have differentiated conductors from insulators, because the free-electron gas should be present only in conductors. This model helps us to understand why there must be almost no difference between the two systems at room temperature, because the E_F energy is found to be several electronvolts at the density of electrons in metals but the thermal energy is a hundred times less; thus almost all the electrons remain in the frozen state. The gas of electrons in a metal would become disordered only at the temperature of the stars; therefore, the electron gas remains ordered at all temperatures where the metal is still a solid and also in a situation nearly similar to absolute zero.

This example shows us that models can be important when dealing with the problem of disorder in the complex structures of matter. We know for certain that electrons in a metal are not free particles, because of the attraction field of positive ions and also because of the repulsion field that operates between them. But further analysis would show that the quantum interaction expressed by the Pauli principle is always the dominant effect, and this interaction is included in the Fermi model.

1.12 Generalization: Disorder and Energy Levels

Our previous models have made systematic use of the system's energy levels and their population, to describe order and disorder in the system itself. At this point, it would be reasonable to wonder whether any quantity besides energy can characterize these aspects of a macroscopic material system. Statistical mechanics shows that we have so far been perfectly correct, thanks to some arguments we shall now mention briefly.

First of all, we must remember that entropy is defined in thermodynamics by means of the heat content, which in turn requires that temperature be introduced as the primary feature for describing the system. Indeed, the very definition of thermodynamic equilibrium is that state in which all parts of the system have the same temperature. But this description of the situation is unsatisfactory, because it artificially separates thermodynamics from mechanics, the science that must describe the properties of matter at every scale. For this reason statistical mechanics starts with the microscopic mechanical properties of matter (such as the quantum states or the equations of motion, depending upon whether we use a quantum or a classical description), and then goes on through probability calculus to establish the statistical distribution of the system's microscopic states.

Of course every microscopic state in the system is characterized by an energy level. An exceptionally high number n of microscopic states is required as the size of the system increases—namely, for every macroscopic body, in the usual meaning of this term. Thus, an extraordinarily large density of levels must cluster around their mean value. It will be easier to convince ourselves of the validity of this statement if we consider a system such as the perfect gas of molecules, where the total energy level is simply the sum of the levels populated by all the molecules, and note that every molecule can exist in any of its levels. Therefore, it is obvious that there are innumerable ways to produce extremely close total energies, and so the macroscopic system must have many levels very close to one another. A rough calculation shows that the separation between levels must be something like a^{-N}, where $a > 1$ and N is the number of particles in the system. Consequently, even in an "isolated" real system, even a weak random interaction with the walls is always greater than the separation among its levels; this implies some inevitable disorder that is truly intrinsic to the system itself, because of the number of microscopic states that the macroscopic system achieves. Thus we must accept this *reality of the intrinsic disorder in a macroscopic system,* and we must realize that this disorder is caused by the existence of many populated energy levels.

The mean disorder in an isolated macroscopic system is thus related to the number of its microscopic states that possess almost equal energy (which we shall call isoenergetic microstates, for brevity). Moreover, note at once that these states are the very ones that are physically distinguishable by means of the energy itself. For instance, we should not consider two microstates that differ from one another merely because two of their atoms are interchanged are different, because there is no operation that can distinguish them; let us recall that quantum mechanics shows that, because of the symmetry of the wave function, a quantum state remains unaltered by the exchange of identical particles. On the other hand, we must realize that states in which two atoms have different energies are truly different, and thus these states contribute to disorder even if the sum of the energies of the two atoms is always constant. Since it is impossible to know the details of the interaction between the different parts of a real system, in practice we cannot calculate the number of these isoenergetic microstates, so it is best to visualize them as the clustering of a very large number of energy levels around the mean value of the macroscopic system's energy.

In this frame, statistical physics introduces entropy as the logarithm of the number n of the occupied isoenergetic microstates,

$$S = k \ln n,$$

where k is Boltzmann's constant, and we merely need to adjust the physical dimensions. Note also that

$$dS = k\frac{dn}{n},$$

and therefore the *increase in disorder, dS, is proportional to the percent increase dn/n in the number of occupied states in an isolated system,* a ratio of direct meaning in a wonderfully simple and precise mathematical expression.

The statistical interpretation of entropy was formulated by Boltzmann in 1877 in the form

$$S = k \ln P,$$

where P is the thermodynamic probability of the macroscopic state in consideration; in other words, the number of ways to achieve this state. Since P is proportional to the number of corresponding microscopic states, and since these all have the same occupation probability, being isoenergetic, it follows that P is proportional to n. It can be easily proved that this so-defined entropy has all the properties predicted by thermodynamics. In particular, if we separate the system into many parts, we can establish that the state of internal equilibrium is the one of greatest probability which makes the total entropy greatest. Note that this very important result is achieved without ever introducing temperature; it relies solely upon the notions of energy and entropy. We should also add in this context that temperature appears as a quantity defined only in terms of energy and entropy, because it can easily be proved that, when several bodies come into contact inside an isolated system in internal equilibrium, one must have

$$\frac{dS_1}{dE_1} = \frac{dS_2}{dE_2} = \cdots = \text{constant} = \frac{1}{T}.$$

Thus absolute temperature certainly characterizes equilibrium, but it is only the common value displayed by this ratio and thus expresses nothing else. Of course, if we wish, we can have a more concrete idea of the temperature T by reconsidering it in the proper energy context. To see this, let's rewrite the above equations in the form

$$\frac{dn}{n} = \frac{dE}{kT}.$$

Then we can interpret the energy kT without using the theorem for the equipartition of kinetic energy. As a result, kT indeed appears as a sort of energy ballast, but it certainly does not have to be seen as a mean kinetic energy. Last, we should add that the notion of temperature as a parameter to measure the distribution spread on levels of different energy (a notion that emerged from our discussion of the spin gas), does not appear in this section because only isoenergetic microstates have been considered here. The ensemble of these last states is called the microcanonical ensemble in statistical mechanics, to distinguish it from the Gibbs canonical ensemble, where the states no longer have the same energy. We have not considered the canonical ensemble here because we have not been trying to offer a summary of statistical physics, but only to justify and generalize the results of the models we have considered in this chapter—namely that the *energy levels are the natural ingredients used to describe disorder in any material system.*

1.13 Summary and Conclusions

In the previous sections we have considered order and disorder in several different models from the viewpoint of statistical thermodynamics, by means of the population distribution among energy levels. Now let's take a quick look at the conclusions we drew from the analysis of each model, and add a few more comments based on their comparison, as seen in Box 1.O.

1. First, it is obvious that *the nature of the energy of the system is irrelevant,* as far as temperature and entropy and thus thermodynamic disorder are concerned. This energy may be entirely kinetic, as in the case of the ordinary monatomic perfect gas; it may be entirely potential, as in the case of the spin gas; and it may be partly kinetic and partly potential, as in the case of the harmonic oscillators of the Einstein model. Moreover, this energy may be either quantum or classical.

2. What truly matters in the case of disorder is that individuals must be distributed in many ways among the system's energy levels. This permits the existence of many microscopic states that are essentially identical in terms of the total energy of the system, but different because of the way individuals are distributed among the various levels and, more generally speaking, among the various states. This is the true concept of disorder. Consequently, the opposite concept of *order should be seen as the correlation displayed by the states of the individuals of the system.* Indeed, at absolute zero only one level must be populated if this correlation is to be greatest, the individuals' states being identical in this case alone.

3. Temperature is seen to be a parameter that characterizes the spread of the statistical distribution on the different energy levels available to the system. Consequently, temperature is a quantity that indicates disorder, because it is inevitable for the correlation between the individual states to decrease as the number of possible energy levels rises, but *temperature cannot be a measure of disorder.* This is particularly clear in the case of the spin gas, where the temperature can even become negative; but a look at Box 1.O shows that this is a general fact, because the functional dependence of entropy upon temperature is different for each of the models we have considered. Furthermore, temperature appears as molecular agitation only in those cases where part of the energy is kinetic, as in the usual case of molecular aggregates.

Box 1.O

COMPARATIVE SURVEY OF THE MODELS

Model	*Form of energy*	*$S = f(T)$ dependency*
Ordinary perfect gas	Classic kinetic translational	$S \approx \ln T$
Fermi gas	Quantized kinetic translational	$S \approx T$
Quasi-particle gas	Quantized vibrational	$S \approx T^3$
Two-level spin gas, infinite temperature	Quantized potential	$S \approx$ cost
Einstein's solid, high temperature	Classical vibrational	$S \approx \ln T$
Einstein's solid, low temperature	Quantized vibrational	$S \approx \frac{1}{T} e^{-1/T}$

Note: We have always omitted proportionality factors in the functional dependence of entropy S upon temperature T.

4. Statistical mechanics defines entropy as the logarithm of the number of the microscopic isoenergetic states of the macroscopic system, and offers a definition of temperature by using energy and entropy. So we have shown that *the disorder introduced to a macroscopic system is proportional to the relative increase in the number of its microscopic states.* Thus it is completely justified to use the statistical distribution of population among energy levels to cope with the problem of disorder in the more general case. The models we have used in this chapter are precisely those that can offer a simple and direct identification of the energy levels.

5. Order in matter must be induced by internal or external forces. However, *order must always compete with the intrinsic disorder of the macroscopic system* because of the many microscopic states displayed by the macroscopic body when in contact with a thermal bath. This competition occurs in terms of energy. The mean energy of the force field and mean energy kT are the two parts in this confrontation. When the system's energy

Box 1.P

FLUCTUATIONS AND CORRELATIONS

- A fluctuation is the displacement Δx of a variable x away from its mean value $\bar{x}$. Naturally, $\overline{\Delta x} = 0$, whereas the mean quadratic displacement $(\overline{\Delta x})^2 \neq 0$. (For further details, see bibliography, [22] or [26].)

- The term thermodynamic fluctuations is used to describe the spontaneous fluctuations of the thermodynamic variables of a system in contact with a thermal bath. For example, $(\overline{\Delta T})^2 = kT^2/C_v$; and $(\overline{\Delta S})^2 = kC_p$.

- The fluctuations of two variables x_1 and x_2 are correlated when $\overline{\Delta x_1 \, \Delta x_2} \neq 0$. For example, $\overline{\Delta S \, \Delta T} = kT$, whereas $\overline{\Delta S \, \Delta P} = 0$.

- The amplitude of a fluctuation generally drops off exponentially in time, and the self-correlation time measures its mean life. Near equilibrium, under the Onsager regime, there are striking symmetries among the time fluctuations of several macroscopic variables.

- The correlation between the density fluctuations in various parts of a fluid was considered in Box 1.L.

is partly kinetic, kT may be expressed as a mean random kinetic energy. In this case, the competition between order and disorder looks like the confrontation of an order-inducing energy, caused by forces that make the system's structure more rigid, and a disorder-inducing energy caused by thermal agitation. These basic ideas make it possible to build a unified picture of a great variety of situations that are generated by forces that can be vastly different.

6. From the statistical thermodynamics viewpoint followed in this chapter, order and disorder are seen to be two complementary concepts, in the sense that the increase of one must imply the decrease of the other. Whereas

order can be reached in only one way (the one corresponding to the closest correlation between the parts that thus merge together), disorder presents itself in quite different forms.

7. The concept of order as the correlation between spontaneous fluctuations will become more and more important throughout this book; therefore, we shall take a first look at it now, in Box 1.P.

1.14 References

(a) The concepts of thermodynamics used in this chapter can be found in every general physics book, as well as in specific works such as Fermi (12) and Zemansky (43).
(b) Statistical mechanics cannot be explained in elementary terms. Perhaps the best treatment of this subject is still the one written by Landau and Lifshitz (26).
(c) The elementary aspects of quantization considered in this chapter can be found in any text on theoretical physics. Born offers an explanation (4) at the elementary level, whereas Slater goes into this subject in greater detail (37).
(d) The concepts of solid-state physics mentioned in this chapter can be found in the very readable book by Dekker (9) or in a more difficult but also more stimulating book by Kittel (23).
(e) The behavior of matter toward absolute zero, the third law of thermodynamics, and negative temperatures are well covered by Wilks (41).
(f) All the topics covered in this chapter are treated in greater depth by Goodstein (15).

CHAPTER TWO

SUPERFLUIDS

2.1 Introduction: Quantum Order and Disorder

Today it is common belief that matter should present the highest degree of order at the lowest temperatures. This correct conviction was formed at the beginning of the century after the statement of the third law of thermodynamics. Nernst initially deduced this law by studying high-temperature chemical equilibria, with no reference to the as yet unknown quantum theory. This law stated that at absolute zero all differences in the entropy values of the various states of a system in equilibrium must vanish. Without this law, entropy at absolute zero would have to increase into infinity if the specific heat remained constant (see Box 1.F). Hence, the third law is satisfied only when specific heat goes to zero with an appropriate temperature-dependent power law. In the previous chapter, we saw that this behavior for specific heat in the case of a solid can be explained only by quantum reasoning, and this leads us to believe that the quantization of energy and the third law of thermodynamics must be closely related.

In reality, although it is of marginal interest in this chapter, it is worth our while to recall that the relationship between the third law of thermodynamics and the quantization of energy is somewhat more general, because only quantization makes it possible to assign an absolute value to entropy and thus to evaluate it in the particular case of absolute zero. This is because in classical statistical mechanics it is shown that entropy is the logarithm of the volume of the phase space accessible to the system, the phase space being formed by the coordinate x and by the $p_x = mv_x$ momentum of every degree of freedom in the system. It takes little time to realize that the absolute measurement of this volume may be taken only insofar as quantization is valid. Indeed, if this volume is divided into small cells,

Heisenberg's uncertainty principle requires that the minimum volume of these cells be defined. Thus the volume of the phase space turns out to be measured directly from the number of cells, which in turn is the number of quantum states of the system. In consequence, *entropy is always a definite number with no arbitrary constants,* and in the particular case when a system occupies only one level, entropy must be equal to zero. This is precisely what happens in every material system as absolute zero is approached, and the lowest energy state has been reached.

In light of the third law of thermodynamics, it might now be believed that matter should show no interesting aspects near absolute zero, because a single-level system seems to indicate a static state of order. But experience has turned up such completely unexpected phenomena as persistent currents of superfluids, which clearly show that the state of order reached does not correspond in the least to the total arrest of all forms of motion, as the classical equipartition of kinetic energy would seem to suggest. Matter retains some residue of motion even at absolute zero, and this is another consequence of quantization, due to what has been called *zero-point energy;* i.e., the leftover energy imposed by the uncertainty principle. For instance, because of this zero-point energy, a harmonic oscillator cannot stand still at its fundamental level, nor can it be simultaneously located in the coordinate x and in the momentum p_x, but it must move randomly around its central position at equilibrium. At low temperatures, this zero-point energy can exceed the thermal energy and thus may manifest itself at macroscopic level in a system of oscillators.

At the lowest temperatures, this quantum disorder may become the dominant disorder, showing up as the random motion of every individual in the system. Quantum mechanics shows once again that this effect increases as the mass of each individual decreases. Thus we are prepared to find greater effects in such aggregates of light molecules as helium atoms, or such light particles as the conducting electrons in metals. But as we shall see, this disorderly individual motion can even generate a state of *overall coherent motion because of the forces acting among individuals.* Indeed, their interaction correlates their movements and therefore produces large-scale effects. Of course, this dynamic correlation is destroyed by thermal noise, except at sufficiently low temperatures. As this chapter shall show us, this overall quantum order can be expressed either as the high occupation of the lowest single level, a concept we understand after the previous chapter, or by further consideration of one macroscopic wave function. The latter development will show itself to be richer and more open to developments in material systems with infinite degrees of freedom.

Let us now try to confine this discussion to the case of simpler molecular aggregates, in which intermolecular forces are essentially central, because they originate from electrons located exclusively on single molecules. Thus it is possible to describe the potential energy between a couple of molecules with only two characteristic quantities, namely the bond energy and the minimum contact distance at zero energy, which in practice coincides with the diameter of the molecule (see Box 2.A). If we include the molecular mass m, these three quantities form a sufficient set for a qualitative analysis of the quantum effects of these molecular aggregates; therefore, let us develop a dimensional analysis, the first step of which will be to form reduced adimensional quantities for some of the relevant thermodynamic functions, which we will indicate with an asterisk. These quantities must have similar values in substances under corresponding conditions. For instance, the critical temperature T_c^* should be constant for all these simple molecular fluids. But this is not the case, and the fact that the deviation is greater as the mass is smaller leads us to think of a quantum correction overlooked by the foregoing analysis. Indeed, let us plot the previous dimensionless quantity as a function of a new dimensionless parameter that characterizes the importance of quantum effects. This parameter may be built as the ratio between de Broglie's wavelength value and molecular diameter. As can be seen in Box 2.A, the reduced critical temperature changes monotonically with the quantum parameter, and the fluid where effects are most marked is helium. Furthermore, the high value of the molar volume V^* would indicate that the atoms tend to stay away from each other because of these quantum effects, and the low value of internal energy E^* would show that a positive contribution is added to the bonding energy. In our convention, bonding energy increases with negative values, because it equals the work necessary to separate molecules from one another. Thus the zero-point energy of the aggregate appears to be a sort of outward pressure, like a kinetic energy that tends to enlarge the material system to allow more dynamic disorder to single individuals.

In this last discussion, the quantum correction of zero-point energy has been introduced as an incoherent phenomenon; i.e., a kind of disorder due to the impossibility of locating independent individuals in the system. In order to understand the possibility that coherent collective effects may arise, we can again make use of Einstein's model, as discussed in the previous chapter, and ask ourselves what difference there is between solid argon and solid helium at the lowest temperatures. We soon realize that, while a solid argon atom may be considered to be well-placed inside its cell, the zero-point energy in solid helium is so high that on the average the helium atom is kept far away from the center of its cell. In other words, in solid helium, the probability of finding an atom in a certain position is a function that is not only widely distributed over the entire cell but also quite appreciable outside the cell. Thus, it is clear that we must be prepared to find these probability distributions superimposed over neighboring cells, to such an extent as to invalidate the separate-cell model of a solid, in which every

Box 2.A

THE QUANTUM PARAMETER OF SOME FLUIDS

Substance	$\mathcal{E}/k$ (°K)	$N\sigma^3$ (cm^3/mole)	Λ^*
He^3	10.22	10.06	3.08
He^4	10.22	10.06	2.67
H_2	37.00	15.12	1.73
D_2	37.00	15.12	1.22
Ne	35.60	12.51	0.59
Ar	119.80	23.79	0.186
Kr	166.10	29.35	0.102
Xe	229.60	36.74	0.064

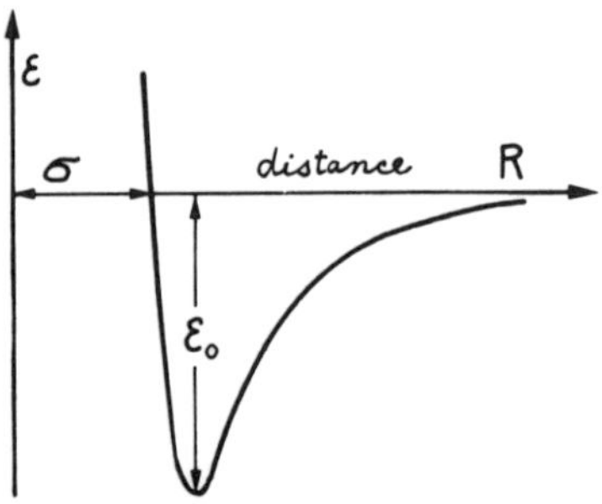

One typical form of the intermolecular potential energy $\mathcal{E}$ (R) for the fluids shown in the above table. N is the number of molecules per mole, while σ is a suitable diameter.

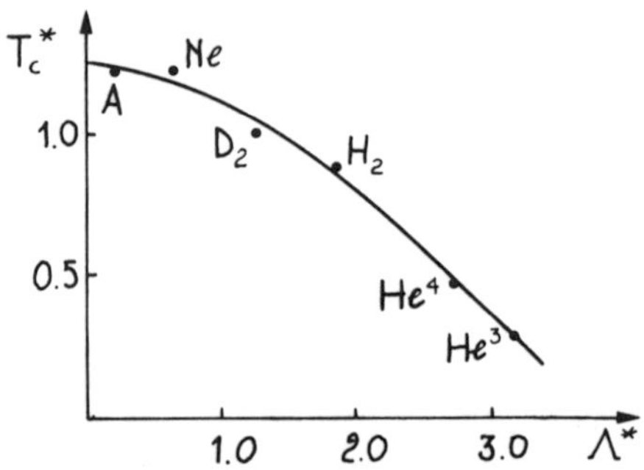

The critical reduced temperature $T^* = kT/\mathcal{E}$ of several fluids of mass m plotted as a function of the quantum parameter $\Lambda^* = h/\sigma(m\mathcal{E})^{1/2}$.

atom moves independently under the constant average field produced by its neighbors. And indeed, at atmospheric pressure, helium is a liquid, never a solid. Thus we must take into account the correlations of zero-point motion in neighboring atoms, and we must begin to think of *the ground level of the aggregate as one single collective level, into which the zero-point levels of the separate atoms merge.* Keeping in mind what we said when considering phonons, which were introduced to give a collective description of the thermal disorder of atomic oscillators, we may now imagine that the fundamental collective state of a macroscopic quantum system can be described by *zero-point phonons,* which obviously do not contribute to the thermal properties, but are nonetheless an intrinsic property of the system.

After this introduction, we are ready to consider the properties of an assembly of helium atoms in a condensed state. We shall see that as absolute zero is approached, a drop of liquid helium must really be thought of as an immense molecule made up of weakly interacting atoms; this molecule, in the fundamental collective state outlined above, must be described in the language of quantum mechanics.

2.2 Liquid Helium: Experimental Aspects

The story of liquid helium is one of an apparently strange liquid which behaves in a manner that directly opposes the norm but which is later recognized to be the simplest and most orderly liquid nature has to offer. The problem it has raised has called for an explanation of how a macroscopic system can be ordered near absolute zero, when that system appears to be so far removed from what common sense would call ordered. As we shall see, we have had to acquire a new way of thinking in order to understand the apparent strangeness of liquid helium and to predict other, still unobserved, strange properties. This new way of thinking was formed to explain the particular case of liquid helium, but it has since been applied to other situations and more complex materials.

Helium stays in a gaseous state down to 4.2 absolute degrees, and it was liquefied for the first time in 1908 at Leyden by Kamerlingh Onnes, who at the same time was also able to lower the temperature to one degree absolute, in the vain attempt to solidify it. Colorless, with low vapor pressure and a low refractive index, liquid helium does not form a meniscus on the glass surface of containers. It was not until 1930 that the first truly important phenomenon regarding liquid helium was discovered. At the temperature of 2.19 degrees absolute, the specific heat took a very curious course, inasmuch as it increased to infinity. This temperature took the name "lambda temperature" because of the form presented by the specific heat as a function of the temperature. While this liquid is absolutely normal above "lambda point," it begins to take on quite a few new properties below that

Box 2.B

PHASE DIAGRAM OF MASS 4 HELIUM

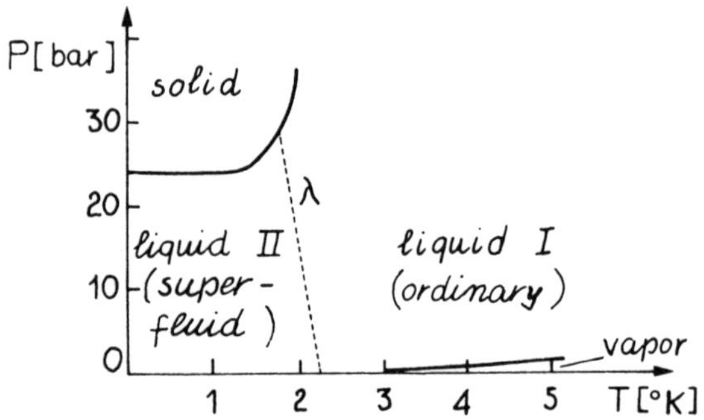

- Note that according to the Clausius–Clapeyron equation

$$\frac{dP}{dT} = \frac{\Delta S}{\Delta V},$$

the superfluid liquid becomes as ordered as the solid, because $\Delta S \to 0$ near 1°K.

- Along the Λ line, an order–disorder transition occurs, with the typical discontinuity in specific heat C_P. The dotted curve is the transition calculated for a perfect Bose–Einstein gas of the same mass as helium 4.

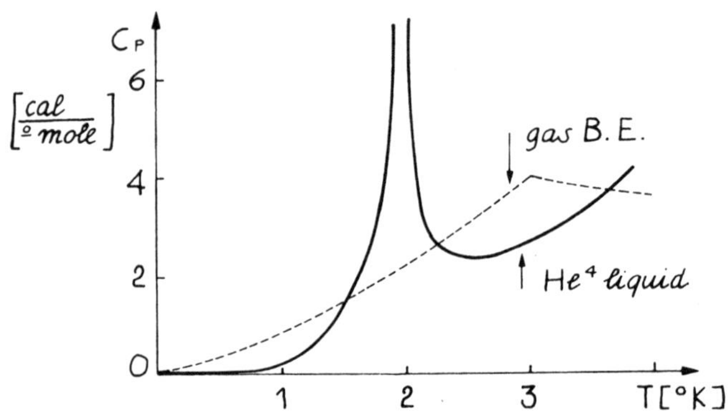

temperature: for instance, it does not boil, even when heat is supplied. The normal liquid at a higher temperature is now called liquid I, while the one below lambda point is called liquid II, thus signifying that the lambda temperature is the transition point between two phases that are both liquid. The first is stable at a temperature above the lambda temperature and perfectly normal, while the second is stable below the lambda temperature and quite unusual, thanks to a series of properties to be described below. This second phase is called *superfluid,* and it will be our main concern in this chapter.

Let us first look at the equilibrium properties summarized in the phase diagram shown in Box 2.B. There is a well-known relationship in thermodynamics, named after Clausius and Clapeyron, which states that the derivative of the coexistence curve between two phases must equal the ratio between the entropy change and volume change in a phase transition. In the case of helium, this ratio goes to zero at lower temperatures along the coexistence curve between solid and liquid II phases. Because there is a volume change between the two states and that change is well defined, it follows that the entropy change must equal zero. Thus we come to the safe conclusion that the liquid must be as ordered as the solid! On the other hand, x-rays that reveal the space order of the structure show that only the solid is crystalline. Thus while the phase diagram shows that the solid and liquid II are equally ordered, the x-rays show that the solid is more ordered than the liquid. Therefore, the inescapable consequence is that liquid II must be ordered, not in the space seen by the x-rays, which is the ordinary space, but amazingly enough, *it must be more ordered in the momentum space,* because entropy is measured in the phase space of statistical mechanics.

Let us go back to the lambda transition. Box 2.B also shows the Bose–Einstein gas transition, because the fact that the order of helium II did not show up in ordinary space but in the space of momenta, suggested in 1939 that liquid helium might be similar to the ideal gas of integer-spin particles. This so-called Bose–Einstein gas has the interesting property (see Box 2.F) of letting its atoms condense into the lowest energy level only below a particular temperature, and when the mass of the helium atom is used, this transition temperature turns out to be very close to the lambda point temperature of liquid helium itself. This was the first suggestion, but Landau later revealed its fallacy.

Let us look at some of liquid helium's other strange properties (see Box 2.C). For instance, there is the disturbing behavior of viscosity, which if measured with a rotating-cylinder viscometer, displays a value as high as water; however, if one measures the helium flow through very fine capillaries, one finds that the viscosity is immeasurably low. Another curious aspect is the occurrence of the so-called helium film on the container walls. If we lower an empty beaker partway into liquid helium, we will soon find the

Box 2.C

SOME PROPERTIES OF SUPERFLUID HELIUM

- The *viscosity* η varies in different measuring devices; the thinner a capillary, the more superfluid passes through it.

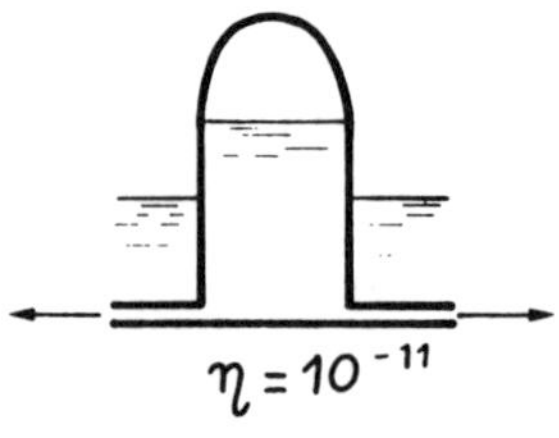

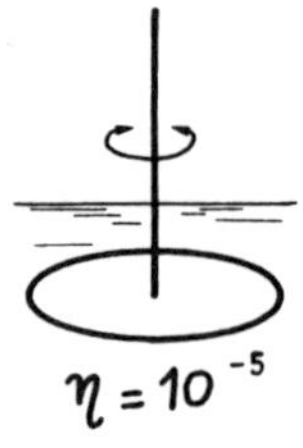

- Cold liquid comes out through a very fine powder and the remaining liquid becomes hotter and hotter, as measured by a resistance thermometer. This device is an *entropic filter*.

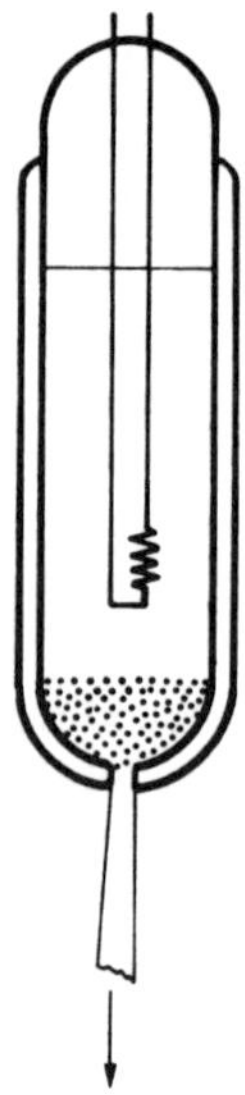

- *Film* flows reversibly without friction along the walls of a glass beaker.

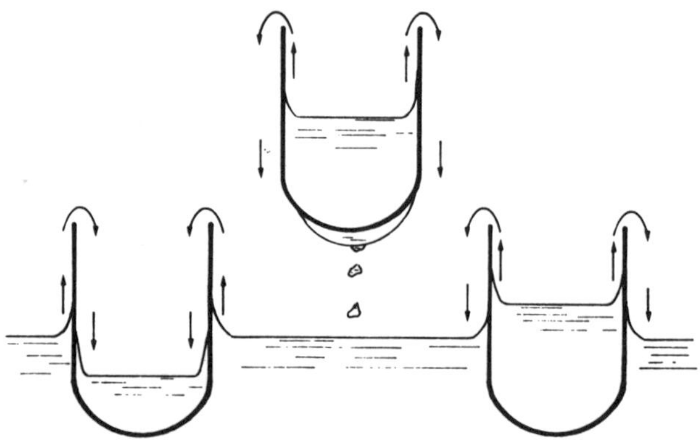

levels inside and out equal; likewise, if we hold a full beaker above the liquid surface, it will empty out after a while. If the beaker is taken out of the liquid (of course, we are talking about a beaker with no holes), it will begin to drip. This can only be explained if there is a film that climbs up the walls of the container, overcomes the force of gravity, and ascends the beaker without friction. However, an even more interesting phenomenon is shown by the "entropic filter." If we take a small container with a hole, then plug the hole with a very fine grained substance (which is usually the extremely fine powder used by jewellers to clean precious stones—this allows us to pack the hole as closely as possible) and then lift this container above the liquid, we are surprised to find that the liquid that remains in the container heats up, while the liquid that comes out is cooler than the original liquid.

All these experiences contained important but always qualitative observations. The quantitative experiment was performed by Kapitza in 1940, when he was able to demonstrate that the cooler liquid coming out of the entropic filter *had exactly zero entropy and zero viscosity,* and therefore it was right to call it superfluid. In the next section, we shall theoretically consider how a liquid can be ordered and superfluid at the same time.

2.3 Liquid Helium: The Two-Fluid Model

On the grounds of information derived from Kapitza's experiments, Landau formulated his revolutionary theory, in which liquid helium was considered a blend of two fluids, the one a perfectly ordered one called the superfluid component and the other, having absolutely normal features, called the normal component. These concepts are of a quantum-mechanical nature and thus must not be considered figuratively: in other words, rather than two liquids which are both present and indistinguishable from each other, these are *two quite different forms of motion which the same liquid is able to possess simultaneously* in all its accessible space. This rather arcane terminology will soon be explained by means of the concepts of collective fundamental level (as outlined in the previous section) for the superfluid, and of quasi-particle gas (similar to the phonons considered in the first chapter) for the normal fluid.

Some examples may make this two-fluid model easier to understand. First, let us try to interpret the experiences of viscosity. According to this model, only the superfluid component must pass through a narrow capillary, while the normal component is stopped by the capillary walls. Thus, measuring viscosity with a capillary should result in a viscosity that is apparently quite low, while the viscosity measurement taken when a large rotor is moved in the liquid should give a different result, because in this case, the superfluid does not interact with the rotor, while the normal component creates friction and gives the impression that the viscosity coefficient is per-

Box 2.D

THE TWO-FLUID MODEL

- Below the Λ point, the liquid is composed of two fluids:
  The *superfluid* component, formed by the population that occupies the collective ground state, displaying $S = 0$ and $\eta = 0$.
  The *normal fluid* component, formed of the quasi-particle gas with this dispersion curve

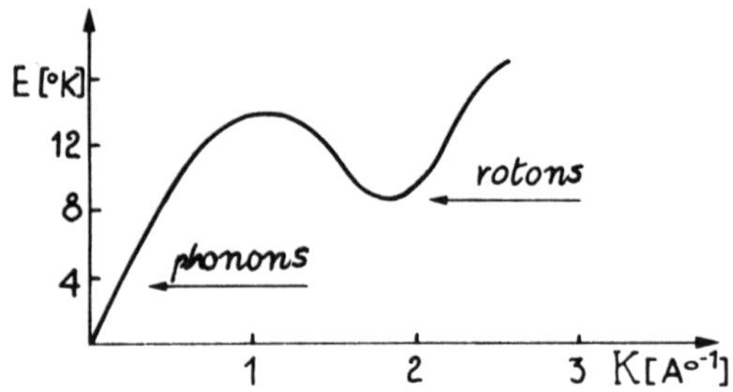

$E = f(K)$
E = energy
K = momentum

- If we call the relative percentages of these two fluids ϱ_s/ϱ and ϱ_n/ϱ, then $\varrho_s + \varrho_n = \varrho$.

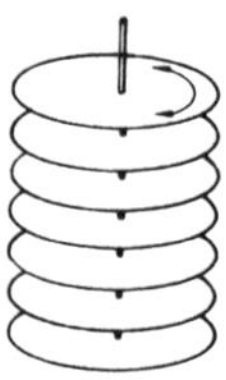

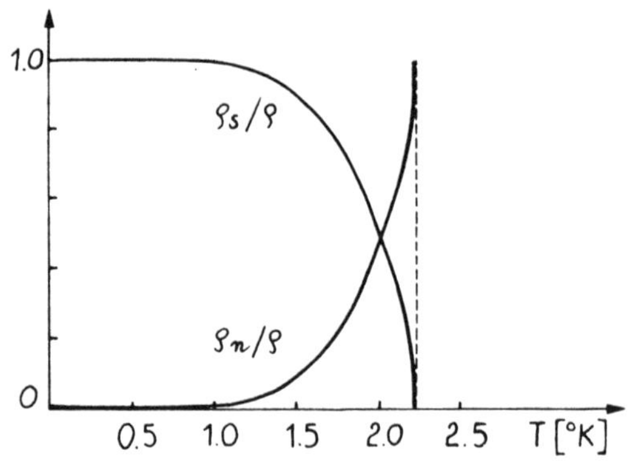

- ϱ_n can be determined using a stack of rotating disks that pulls along the quasi-particle gas but not the superfluid; the apparent value of the rotor's moment of inertia gives the experimental value of their percentages as a function of the temperature. In this way, it has been found that near 1°K, helium is a nearly perfectly ordered liquid.

fectly normal. Moreover, only the superfluid component can go through the narrow gaps of the entropic filter, and because the superfluid component must be thought of as a fluid at virtually absolute zero, inasmuch as it is totally free of entropy, the jet we observe should be cooler because it consists of the superfluid component alone.

In Landau's model, the superfluid component is made up, not of atoms, but of the *collective fundamental level* outlined in the previous section; as there is only one level, it has zero entropy and cannot exchange energy. Thus the superfluid remains isolated from the environment because it does not exchange energy with the walls and its viscosity is rigorously zero. All *the disorder is in the upper levels,* which are populated at finite temperatures and give rise to the normal component because these excited levels, thought to be a quasi-particle gas, exchange momentum when they hit the walls, thus causing viscosity. So the conceptual point of view is a completely new one. Whereas before we thought in terms of isolated atoms, it is now the whole beaker of liquid helium that possesses both these natures at once, because we must not think that an atom can be in one fluid part of the time and in another the rest of the time: the truth is that all the atoms participate together in the formation of all the collective levels.

Landau went further. He proposed experiments that led to total triumph of his point of view. For instance, with reference to Box 2.D, let us take a pile of disks as our viscometer. If these disks are stacked tightly, we may obtain a distance at which the quasi-particle gas is trapped between them as they rotate, while the atoms can certainly escape. In other words, a tight enough stack takes the quasi-particle gas along with it if the gas has a mean free path long enough to hit the walls. Now, if the stack turns, the moment of inertia varies as a function of the temperature, because when it is close to lambda point the stack takes almost the entire mass of liquid with it, while at lower temperatures it takes almost nothing along. It is evident that this experiment is crucial because it cannot be interpreted in terms of isolated atoms but only by postulating the presence of a quasi-particle gas with a density that increases with temperature.

Landau was also able to predict other extremely interesting phenomena that led to clearer and clearer evidence of these quasi-particles. For example, just as the oscillating motion of atoms in space leads to the phenomenon of sound, the oscillating motion of quasi-particles must lead to new phenomena, which Landau called "second sound." These new periodic phenomena show up at the macroscopic level as a temperature wave-distribution, because, as in sound, the density of matter is distributed according to the laws of wave motion; now second sound must also be thought of as the local compression and later rarefaction of the quasi-particle gas. The increased density of the quasi-particle gas leads to a rise in temperature, energy, and entropy, while its rarefaction must lead to a local increase of the superfluid component and thus to the opposite effects. The most interesting thing is that these phenomena obey the laws of wave motion, which means that the temperature in a channel full of helium fluctuates in a wave pattern, just as do sound waves in an organ pipe. This can be effectively revealed, in the case of second sound, by moving a small thermometer along the tube where the wave propagation is occurring; we will then see the maxima and minima of temperature, which are analogous to the peaks and valleys of density in the acoustic phenomena.

At later stages, the two-fluid theory was subjected to a stringent microscopic test by introducing ions into liquid helium and then moving them by means of an electric field. It is clear that slow ions can interact only with the normal component and not with the superfluid one. Thus it is possible to measure the drift of this type of charged sphere and thereby determine the microscopic processes of collision between ions and quasi-particles of even microscopic size. According to the two-fluid model, when temperature goes down so does the number of quasi-particles, and thus there should be a corresponding increase in the mean free path of an ion, just as there is an increase in the mean free path of an ion in an ordinary gas when the pressure of that gas is lowered. And in fact, in liquid helium at 0.3°K, the ions have a mean free path 10^5 times greater than that found at lambda point. Near absolute zero, when the normal component vanishes, the only leftover disorder that limits the ions' path is the one caused by the He^3 impurities inevitably present in natural helium. Furthermore, the overall motion of quasi-particles can be followed by mixing ions into the normal fluid, just as a drop of ink can be used to follow the motion of a fluid in a pipe. This method made it possible to perform a more rigorous test of the hydrodynamic equations of the two fluids, and thus irrotational phenomena were discovered that Landau had not predicted. We shall consider these phenomena in the next section.

Landau was not able to find a microscopic theoretical justification for these quasi-particles, which represent the collective energy levels of liquid helium, with the empirical dispersion curve seen in Box 2.D; the curve is a plot of energy versus momentum, and thus for a free particle would be a parabola. At first, this curve was postulated *ad hoc,* but it was fully verified much later with the neutron technique we saw in the previous chapter when dealing with spin waves. In that case, neutrons were of interest as spin carriers, whereas in these experiments they are mass carriers able to exchange momentum and kinetic energy with the beaker full of liquid helium. In this dispersion curve, there is an initial linear region of phonons followed by a parabolic region with quasi-particles that Landau called "rotons," because he made an attempt to quantize rotational motion in hydrodynamics. As we know, a phonon arises as a wave-type fluctuation of density, and is the quantum of the elastic field. But Landau never was able to solve the mystery of the roton. As we shall see below, Feynman thought it was the smallest vortex ring that can exist, a correlation of the fluctuations in the velocity field. With the neutron technique, the line width can be used to calculate the mean life of these quasi-particles, and it is found that when lambda point is reached, rotons have a very low mean life. In other words, the quasi-particle gas now becomes a dense fluid, and lambda transition may be thought of as a cooperative effect in this ensemble of quasi-particles. Thus, near lambda point the two-fluid model is proved to be an oversimplification, and the concept of quasi-particles loses its usefulness.

2.4 Liquid Helium: Vorticity Quantization

It did not take long to discover the limits of Landau's theory. Experiments carried out with improved accuracy showed that the superfluidity of liquid helium moving through a capillary broke when the liquid's critical velocity exceeded its critical velocity of a few centimeters per second, whereas Landau's theory predicted that this should occur at about sixty meters per second. Indeed, Landau postulated that superfluidity should be broken by slowing the fluid down while creating a quasi-particle at the same time. It can easily be seen that if a quasi-particle is to be created, the fluid's velocity must be close to the slope of the tangent to the dispersion curve displayed by the energy spectrum shown in Box 2.D, and thus must be close to the speed of sound.

In 1954, Feynman proposed to explain such low critical velocities by introducing a somewhat new element into play: that of quantized vorticity. Just as classical hydrodynamics proves that the velocity circulation around a vortical line is constant, likewise for a quantum liquid obeying Bose–Einstein statistics, Feynman said that this constant had to be an integer multiple of h/m, where h is Planck's constant and m is the mass of a helium atom. Thus the fluid elements in motion around a vortical line must be subject to a condition resembling the one Bohr applied to electrons rotating around the nucleus. Onsager had already mentioned this possibility at a conference in Florence in 1949, but his remarks had gone unnoticed, perhaps because of this great theoretician's notorious abstruseness. To return to the excessively low critical velocities at which the flow of superfluid in a channel is broken, Feynman was able to show that this phenomenon might in fact be due to the creation of quantized vortices, because of their peculiar energy-versus-momentum dependence.

The most direct way to visualize the creation of vortices in a dissipative process can be obtained by moving an ion through the liquid at sufficiently high velocities, using the technique mentioned in the last section. Just as a sphere moving through an ordinary liquid leaves a wake of vorticity, an ion accelerated through liquid helium by an electric field must also produce a wake made of single elements of quantized vorticity. At low

Box 2.E

QUANTIZED VORTICES IN SUPERFLUID HELIUM

- Remember that in the rotation of a classic fluid the surface assumes the shape of a paraboloid; whereas, if a vortex forms, the surface becomes an internally empty funnel.

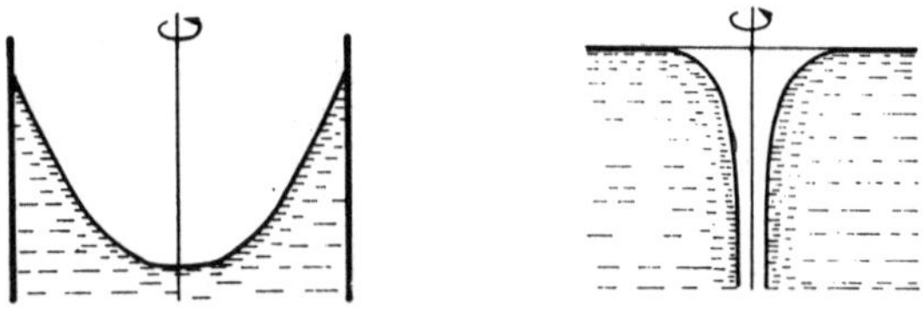

- In 1949, Onsager postulated that the circulation of superfluid helium around a vortical line could be quantized, as for the motion of an electron around its nucleus.
- In 1955, Feynman proposed that in the rotation of a container full of superfluid helium, some quantized vortical lines would lie parallel to the axis of rotation

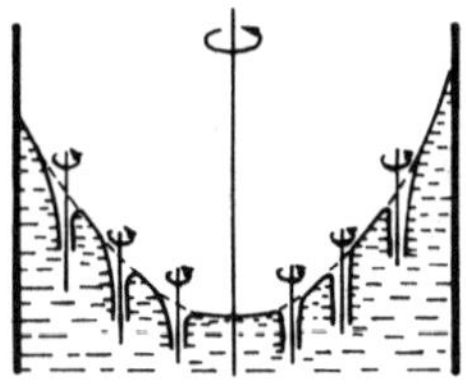

- Circular quantized vortices can also take place; analogous to ordinary smoke rings, their motion can be detected by trapping an electron inside.

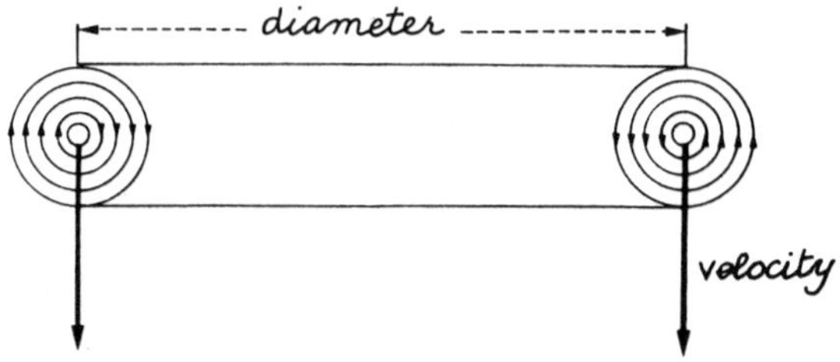

enough temperatures, when the superfluid is almost the only component, a single circular quantized vortex ring can be accommodated around the ion. Of course, the need to quantize circulation leads to the phenomenon of resistance to motion differing slightly from classical hydrodynamics.

Macroscopic quantized vortices were discovered in 1956 for the case of an easily theorizable situation, namely a rotating cylinder filled with superfluid helium. In this case, Feynman predicted vortical lines parallel to the cylinder's axis of rotation, as shown in Box 2.E, and experiments did in fact show that the absorption of second sound was not isotopic when the cylinder was rotating; this indicated the presence of a definite number of vortical lines. However, the details of this interaction between vortical lines and second sound are far from simple. A more intuitive way to establish the presence of quantized vortical lines is to trap electrons in the empty region that should exist at the center of the vortical line. Electrons can be trapped this way because their low mass gives them a high zero-point energy that can be lowered if they accommodate themselves in a region wider than the one between atoms in a condensed phase. In this way, it is even possible to deal with macroscopic isolated vortical lines containing a few trapped electrons, and to determine their quantized mechanical properties by applying external electrical fields.

Once the quantization of circulation has been established at the macroscopic level, it is plausible to imagine that there should be a vortical ring of such small microscopic size that only one atom can pass through it, and that this elementary vortex may be the roton postulated by Landau. Thus, like the other quasi-particles we have already mentioned, this element of disorder does have its own structure, because it can be visualized as a correlated local fluctuation of the hydrodynamic velocity field. At this point, we can go further, using this roton image to propose what may occur at the lambda point. Inasmuch as the roton number greatly increases, it may be expected that this transition is due to the interaction among this very large number of rotons via their velocity field. Indeed, Feynman made another proposal worthy of mention, despite its lack of experimental verification: he said that all the rotons at lambda point may merge into one single vortical line dense enough to fill the entire space occupied by the fluid. In other words, a long-range correlation might be formed in the velocity field throughout the fluid. And in fact, as we shall see in the next chapter, recent theories on phase transitions have shown that at the critical point some correlations between fluctuations must extend to an infinitely long range. Thus Feynman's suggestion of the lambda point may be correct.

Although we have limited ourselves to a qualitative description of the properties of liquid helium, it should nonetheless be clear that we are dealing with order and disorder that are both quantized. Above the ground level that implies a collective kind of order, there is a branch of energy levels that fit the Landau quasi-particles, but there is also another branch of energy levels that are the Onsager and Feynman unstable vortices. A vortex always implies a correlated motion between the elements of a wide fluid region, and in this case the quantization of circulation makes the correlation even stronger. The presence of these quantized vortices of even macroscopic dimensions is obvious proof that *observation of quantization is not dependent upon space dimensions;* instead, it depends entirely upon quantum energy, and thus quantization can always be observed when the thermal noise of the bath is below one quantum of energy. But the real importance of the quantization of circulation seen in superfluids was to call attention toward the phase of the wave function that describes the ground collective state, thanks to a simple analytical development (omitted here) that relates this phase to the circulation around a vortical line. The description given by statistical thermodynamics, which states that maximum order is to be found only when one single energy level in the system is occupied, no longer suffices to explain all the various properties displayed by superfluids. *Order in superfluids must be expressed using the coherence properties of a wave function.* This conceptual development is no less important than the introduction of quasi-particles, a new development whose examination in the next sections will lead us to the order-parameter notion considered in the third chapter. So the ground level of liquid helium becomes similar to the "vacuum" of quantum electrodynamics, a peculiar structure with infinite degrees of freedom that has played a major role in the physics of elementary particles.

2.5 Bose and Fermi Liquids

In the sections above we briefly mentioned the possible importance of Bose–Einstein's quantum statistics for the behavior of mass 4 liquid helium, also simply called liquid helium because it is the most abundant isotope in nature. We gave little notice to the importance of quantum statistics, in order to concentrate on other aspects of the problem; but now we shall return to it, because of the vastly different behavior shown by the two helium isotopes in the condensed state in the same temperature range. Since the same intermolecular forces in both cases must originate from the same number of electrons, mass 3 helium behaves differently from mass 4 helium because of the different quantum statistics called into play by the two different nuclei. Thus, this is a clear example of the influence that the symmetry properties of the microscopic components have on the macroscopic behavior of a material system. But, as we shall now go on to see, the role played by quantum statistics is not as simple as it might be if the two liquids could be described as the two perfect gases, Bose's He^4 and Fermi's He^3. Actually, it is the study of liquid helium that has shown that we must think in terms

Box 2.F

BOSON-GAS CONDENSATION

- In a perfect boson gas, the entropy departs from the classic value below T_c; the ground level of zero kinetic energy begins to fill up, and thus these particles must have zero velocity.

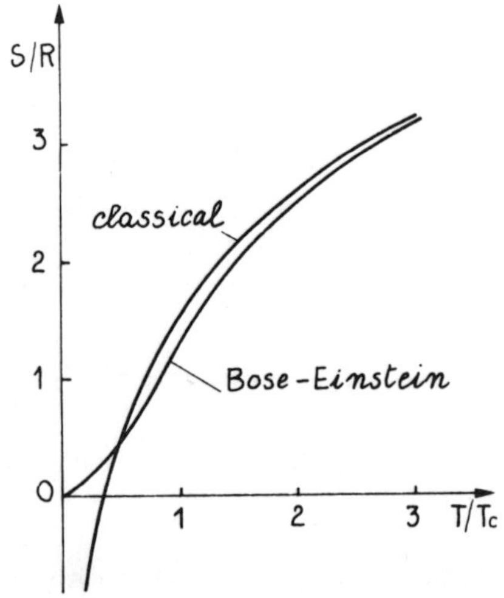

- Below T_c the occupation n_0 of the ground level brings to mind the magnetization of ferromagnets.

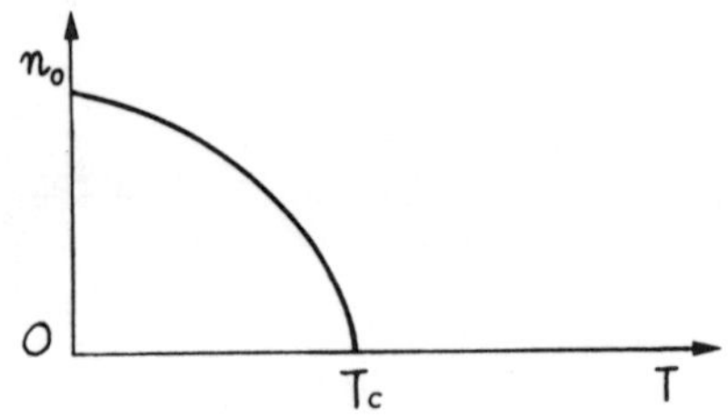

- In this case we are dealing with two fluids made up of atoms of different velocities, whereas in the Landau model the two fluids are formed by the system's collective levels.

- Unlike any ordinary fluid, which can be localized in ordinary space, in the boson gas there is a low-temperature order in the velocity space only.

Box 2.G

PHASE DIAGRAM OF MASS 3 HELIUM

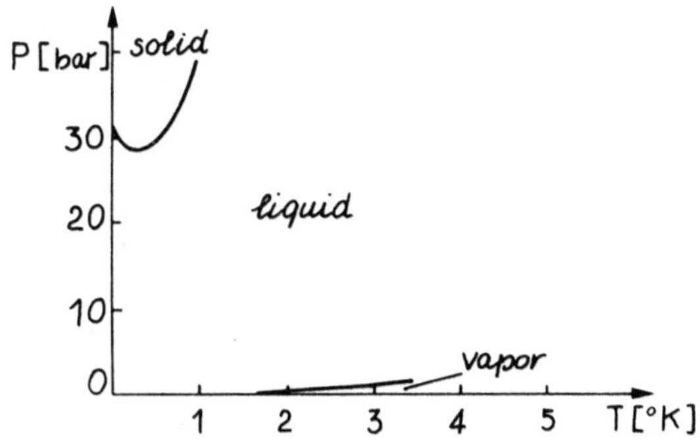

- Note the vast difference between this figure and the one shown in Box 2.B for mass 4 helium.

- Further experimental developments in the extremely low-temperature region have led to the discovery of two superfluid phases.

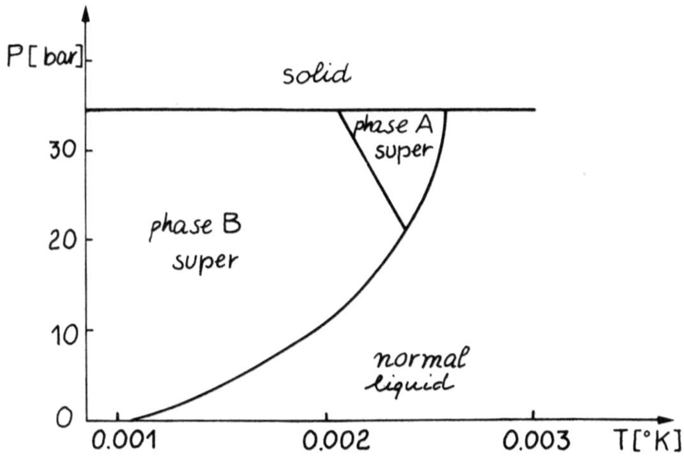

not of isolated atoms but of collective energy levels, yet we know that only isolated atoms are considered in the perfect gas model. The Bose–Einstein condensation certainly means an increase of the ground-level population below a critical temperature, but this is still an ensemble of isolated atoms and not a fundamental collective level generated by proper overlapping of the wave functions of many interacting atoms. For this reason, we have ruled out the interpretation of the lambda point as the transition of a perfect Bose–Einstein gas.

A more correct view of the clear influence of quantum statistics can be obtained by proper consideration of the collective energy-level spectrum of the two liquids with different masses. In the case of mass 4 liquid helium, superfluidity is a direct consequence of the form of the spectrum shown in Box 2.D, because of Landau's argument mentioned briefly above. If they are to be created, even one at a time, the quasi-particles in this kind of spectrum must have an integer spin, so that the angular momentum of the liquid helium in the beaker can assume quantized values. The details of the spectrum are hard to predict theoretically, but this is only of secondary importance; besides, this always occurs for a system with N bodies, when N is close to Avogadro's number. Incidentally, we should point out here that Feynman's derivation of this spectrum is a qualitative one, but it is based on those quantum-mechanical symmetry properties upon which Bose–Einstein statistics is founded. Likewise, these same properties are also used to introduce quantized vorticity.

The spectrum of mass 3 liquid helium is vastly different, and it is precisely the nature of this spectrum that makes it impossible to observe superfluidity in the same temperature range in which it is observed in mass 4 liquid helium. Actually, *the spectrum of a Fermi liquid should not differ greatly from that of normal metal electrons* mentioned in the first chapter, except that Fermi's energy level should be much lower because the mass of He^3 is much greater than the mass of the electron. According to this oversimplified line of thought, we should expect the Fermi liquid made up of He^3 at low temperatures to show somewhat unusual behavior because of its high zero-point energy, as shown in Box 2.G, but there should certainly be no superfluidity.

What actually happens is that the mass 3 liquid helium becomes superfluid below a few thousandths of a degree, exactly because at this temperature its energy spectrum undergoes marked change. As a matter of fact, so far we have not considered the magnetic interaction that exists between He^3 nuclei, and it is just this weak energy that now allows atoms to pair off at such a low temperature. *From the viewpoint of statistics, the Fermi pair behaves like the one single boson, and the energy spectrum is accordingly modified and allows superfluidity.* This fact is similar to the pairing of electrons which leads to superconductivity, but the interaction that results in pairing electrons in superconducting metals has a different origin. The spectra of these two boson fluids made up of Fermi pairs are indeed quite analogous. In both cases, there is dense occupation in a state kept separate from all the other higher-energy states, and this dense state plays the same role as the one played by the ground collective state in the case of He^4.

Of course, in the case of He^3 the different ways of spin pairing must give rise to different situations, and in fact two superfluid phases have been found in the laboratory, whereas in the presence of the magnetic field even three phases have been observed. At the macroscopic level, these superfluid phases are characterized by a marked magnetic anisotropy because of the peculiar order imposed by quantization, which cannot be intuitive. For instance, theory states that the lowest energy state of helium 3 should be that of a liquid in permanent rotation in a container at rest. Such properties have been observed for both mass 4 liquid helium and superconducting electrons, but in these cases they are caused by the occupation of higher energy levels such as the vortical levels. Any theory that predicts a macroscopic object should even rotate at the ground state contradicts our current image of order as an absolute lack of motion, and this comes from our habit of observing the behavior of macroscopic objects around us that obey the classical mechanics laws. The experimental study of liquid helium 3 at such low temperatures has just begun, and because it can offer so many more situations than those offered by mass 4 liquid helium, it will certainly become the most natural testing ground for developing the proper idea of quantum order. On the other hand, superconducting electrons are an ordered Fermi liquid that has already been studied widely, and thus we would do better, in the next section, to discuss the wave function of this condensed boson state. As we have already remarked for the quantization of circulation, the properties of this wave function can offer a better description of the superfluids than the one we can deduce from considering only their energy spectrum, as we have done in this section.

2.6 Superconducting Electrons: Macroscopic Aspects

At low temperatures, many metals drastically change those properties that involve conducting electrons, for instance, the unexpected and complete disappearance of resistance. Because this phenomenon is quite apparent, it is called the superconductive transition, even if a closer look shows that it is the consequence of an exceptionally high diamagnetism, a fact that we shall next consider. It was mentioned in the previous section that these strange properties are caused by the formation of electron pairs behaving like true bosons, but we shall return to a interpretation of superconductivity at the microscopic level in the next section. Here we want to see that a state of order originates in these metals at a macroscopic scale, and that this state of affairs can be described by a collective wave function. Experimentally, this order becomes amazingly evident for the intrinsic electric and magnetic properties displayed by the electrons.

First, let us consider the behavior of the entropy of a metal when it becomes a superconductor. Without an applied magnetic field, entropy drops sharply when the temperature is lowered below a certain critical temperature, in perfect analogy with what occurs in ordinary liquid helium, as seen in Box 2.H. It is clear that this process must involve only the free electrons of the metal because this entropy variation comes from the electronic specific heat alone and not from the lattice contribution accounted for in Debye's law. But unlike the order of liquid helium, this order can now be destroyed if an external magnetic field exceeds a value which is also critical. Thus we are forced to consider the magnetic properties of the superconducting state.

The magnetic properties are quite unusual in the superconducting state, and they may be summed up by saying that the lines of the external magnetic field are pushed out of the space occupied by the metal, this being the so-called *Meissner effect* shown in Box 2.H. The reduction of the external magnetic field inside a piece of matter is a general property known as diamagnetism, and it is caused by the reaction of the atomic electrons to the disturbance imposed on their Bohr orbit. More precisely, some atomic currents are generated and they produce an additional magnetic field that tends to reduce the disturbing effect of the external field. Normal diamagnetism is a general property of matter, but its value is very low and therefore is not of importance here, whereas for superconducting metals diamagnetism is almost infinite. A superconducting metal becomes a truly perfect diamagnet in the temperature range and in the magnetic-field intensity at which the superconductive phase is stable. Using Maxwell's equation of classical electrodynamics, it can be shown that a perfect diamagnet must also possess exactly zero ohm resistance, and therefore the superconductivity of these metals is actually only a consequence of their infinite diamagnetism. So here we shall limit ourselves to considering this anomalous diamagnetic response as the basic electromagnetic property of the ordered phase established in the free electrons when entropy drops. To continue the analogy with the above-mentioned case of an isolated atom, we may say that the Meissner effect must mean that there is a united response from all the electrons sitting in the conduction orbitals. Thus these electrons manage to generate some nondissipative electrical currents on the surface of the body that are capable of completely excluding the external magnetic field, and of course this all holds true no matter the shape of the superconductor. The way these nondissipative currents are generated spontaneously all over the macroscopic body, regardless of its shape, is an awesome and true manifestation of *the presence of a long-range order in the motion of the free electrons*. Inasmuch as the motion of free electrons in a normal metal is well described by quan-

Box 2.H

SOME PROPERTIES OF SUPERCONDUCTORS

- The entropy S in the superfluid state is lower than in the normal state, which can be restored by applying to the superconductor a magnetic field larger than a certain critical value.

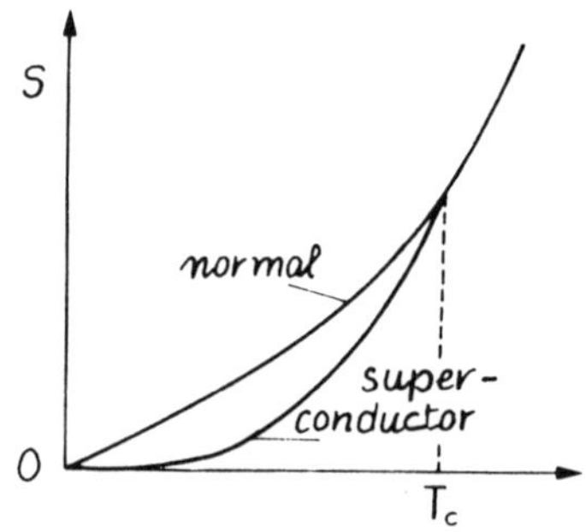

- Below the critical field, the A ↔ B transition is reversible, and the lines of force of the external magnetic field H do not penetrate the superconductor because of its perfect diamagnetism.

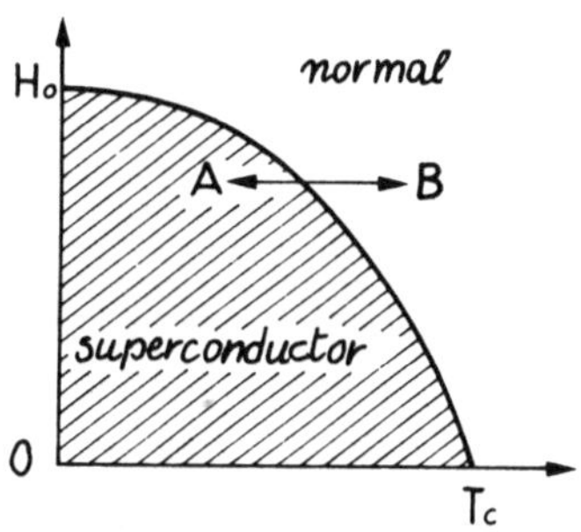

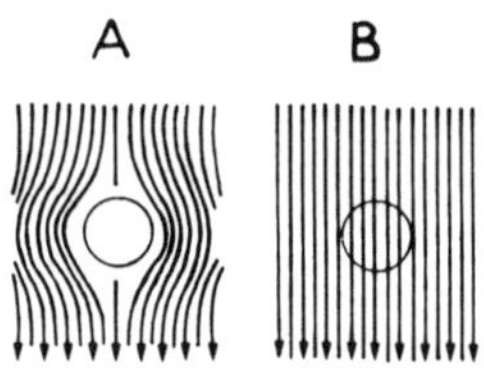

- Several experiments on the tunneling effect in a normal superconductor–insulator metal junction show that the energy spectrum of superconductor metal electrons is greatly modified, giving rise to a gap Δ near the Fermi energy E_F.

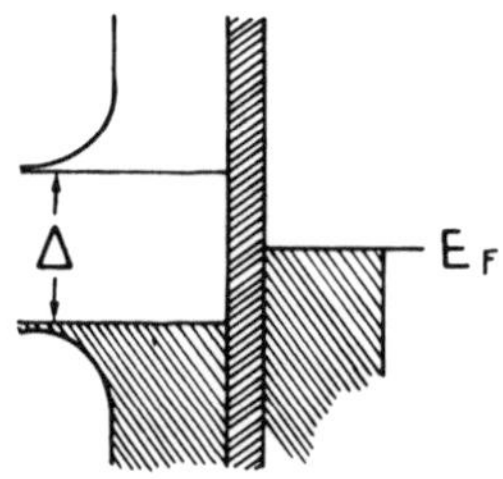

tum mechanics in terms of probability waves, each associated to an electron by the well-known de Broglie relation, the perfect diamagnetism of the superconducting state can be seen as the formation of one single probability wave for all the electrons in motion. Thus there must be one single wave function that encompasses the macroscopic body and that manifests its rigidity toward any outer disturbance. We should concentrate on this important fact, which calls for further comment on the meaning of this wave function.

It is well known that, when we are dealing with a single particle, the wave function of Schroedinger's equation gives the probability of finding the particle in a space point, because the particle cannot be localized in only one point. Naturally, when we are dealing with only one electron we must not think that its charge is found in space according to a probability distribution, because the entire charge of the electron is always concentrated in one point in any moment, and it is precisely this probability distribution that Schroedinger's equation can offer. But the outlook changes drastically if instead of one particle we have a large number of them and they are all located in the same energy state with the same wave function. In this case, we can give different interpretation to wave function, because the amplitude of each single probability distribution summed up with all the others gives the real charge density that is distributed in space. This is also the reason why a group of identical photons in a cavity gives rise to one single monochromatic electromagnetic wave, and this is possible because the photons are bosons and tend for statistical reasons to sit in the same state. But ordinary electrons must behave quite differently, because they are Fermi particles and can never be in the same state, according to Pauli's principle. Conversely, it follows that since the electrons that take part in the superconducting state display one wave function, they must behave like boson pairs, a possibility that we discussed in the above section.

Therefore, the *macroscopic charge distribution of superconducting electrons can be described as one single analytic function of complex variables* like the one given by Schroedinger's equation, and this leads to many consequences that can be experimentally verified. For example, the complex nature of this analytical function implies that the circulation taken around one point in a multiply connected field, such as an empty hole in a plane, must be constant and quantized. This immediately leads to the quantization of circulation around a vortical line, which we have already seen for liquid helium, and which must now hold true for the fluid made up of superconducting electron pairs. Thus, according to Ampère's theorem, even the magnetic flux through a metallic superconducting coil must be quantized. At its roots, this is identical to the quantization of vorticity displayed by liquid helium, but now this is more striking because it is possible to directly measure the magnetic effect associated with the supercurrent circulating in the coil.

Josephson predicted some interesting effects caused by the presence of this phase in the complex nature of the wave function. For instance, one can obtain the interference of two superconducting currents in the same way one can obtain the interference between two light beams of the same frequency and phase. Naturally, the two superconducting currents can interfere only through an insulating junction, so that the phenomenon can be seen as the tunnel effect of boson pairs that are all in the same wave state, just as in the optical case it occurs to photons coherently produced by the same source of light.

In 1954, London proposed that the unusual properties of both liquid helium and superconductors could arise from a quantum-mechanical macroscopic feature, to be identified in the formation of a single long-range-order state of the momentum vector. But this hypothesis did not become a clear reality until the quantization of circulation was expressed in terms of a collective boson-wave function. Thus it is now certain that the quantum order of a superfluid must be described using two parameters—the amplitude and phase of the wave—which together characterize the space–time coherence of the wave function. This wave function will be considered as an *order parameter* to describe the behavior of these systems in the order–disorder transition that leads to the formation of these superfluid states, within the more general framework of the phase transitions later developed, to be seen in the next chapter.

After all we have seen in this section we can suggest that the macroscopic order presented by superconductors is not unlike that of superfluid helium, except for those aspects of the former fluids that are related to the presence of electric charges. On the other hand, superconductivity is not a general occurrence that always takes place in a Fermi fluid, and this makes it necessary to consider its microscopic nature, which we shall do in the section to follow.

2.7 Superconducting Electrons: Microscopic Features

As we know, superconductivity is quite easy to observe in the laboratory. This might give it the appearance of being a simple property, thus easily describable in microscopic terms. But this has not been the case. The correct theory, called BCS in honor of its three authors (Bardeen, Cooper, and Schrieffer), came more than half a century after the discovery of superconductivity much earlier (in 1908), and it is anything but elementary. In this section, we shall try to describe the microscopic state of order reached by superconducting electrons, in light of this theory but without going into quantitative details. For simplicity's sake, let's start at absolute zero and ask ourselves how electrons can manage to form an ordered state that differs from the normal Fermi–Dirac distribution, which we know has zero entropy at absolute zero as well. Actually, this question is misstated, because electrons in metals are not a perfect Fermi–Dirac gas, thanks to the forces still at work among them. As we shall see, these rather complex forces involve the crystal lattice where the free metal electrons move, and it is for this reason that superconductivity is related to the particular nature of the metal in question, instead of being a general feature of the Fermi sea alone.

Thus the situation of superconducting electrons brings to mind that of superfluid helium of mass 3, but the former is more complicated at the microscopic level because the additional forces at work among fermions are more complex. In the case of helium 3, we saw that these forces were the weak magnetic interactions between atomic nuclei. In the case of superconducting electrons, these additional forces result from the *dynamic deformation of the crystal lattice induced by the electrons in motion.* This effect gives rise to a weak reciprocal attraction between two electrons, because to reduce the energy they try to be together in a region of the medium that is denser and therefore more easily polarized. This interaction, which was first proposed by Fröhlich, calls for an extra energy that is a great deal weaker than the kinetic energy of free electrons by a factor of about 10^{-8}, but even this slight contribution to attraction is enough to couple these electrons into so-called Cooper pairs, thus transforming two electrons with antiparallel spin into a zero-spin boson. Further consideration shows that the total angular momentum of most of these pairs is equal to zero. Thus the ensemble formed by this group of pairs is highly correlated, because the vectorial values of the electronic momenta and spins are equal and opposite.

Now let us try to understand the energy spectrum of the electrons in a superconductor at absolute zero. We shall start with the principle mentioned in Chapter One, which states that only the electrons close to Fermi energy E_F can change their state because they are the only ones close to the empty upper levels where they can go, whereas the other electrons, in Fermi's sea below E_F, are always blocked at their energy levels. Then the energy spectrum of a superconductor must differ from the spectrum of a normal conducting metal only near the E_F energy, where the additional force among electrons in motion can operate. Inasmuch as this force works by forming bosons, and since for statistical reasons bosons tend to stay in the same state of lower energy, the energy spectrum of the superconductor should basically be formed by *a single highly populated level containing all the boson pairs* located below the empty levels of the single-particle, Fermi–Dirac gas distribution. In order to form this single highly occupied energy level, we must take electrons away from the population close to E_F. This can be done by taking them from the previous distribution of the perfect Fermi–Dirac gas, thus creating an empty region of levels above the one occupied by the pairs. So a region of forbidden levels, commonly called the energy gap, is formed as shown in Box 2.H. This kind of spectrum is reminiscent of the semiconductor spectrum, but in that case the gap is much greater and is caused by the distance separating valence and conduction bands, and thus for a completely different reason.

The shape of the energy spectrum of a superconductor shows up in various ways in the laboratory. The easiest and most direct way to visualize this gap above the highly occupied boson-pair level is to study the junction formed by a superconductor and a normal conductor. Bearing in mind the electrical conduction behavior of a potential barrier, we can imagine that the value of the gap can be directly derived by observing the behavior of the current flowing through the junction under applied potential.

Let us go back to the boson pairs. These must be understood to be made of electrons whose particular spin and momentum value causes them to attract one another with an energy close to that of the gap. These electrons can then become uncoupled if they absorb such energy. However, we should not think that each of these pairs is formed of two well-defined, stable, joined, or even neighboring, electrons. *The pair concept is a statistical one,* because the average distance between two electrons in the pair encompasses about 10^{-4} cm, a distance within which there are many other pairs. The statistical correlation between two electrons in the pair thus extends to a macroscopic mean value called the *coherence distance.* This quantity can be experimentally evaluated on the basis of the distance separating the superconducting part from the normal one, a figure we can obtain by applying an external magnetic field that destroys superconductivity only in this region of the metal.

Finally, let us mention that the superconductive transition also can be seen as the cooperative break of the boson pairs, and that the energy gap actually does go toward zero at the transition temperature. We shall not consider this superconductive transition further, but this phase transition will appear again in the next chapter, where we shall also deal with the other order–disorder transitions.

2.8 Summary and Conclusions

In this chapter we have considered several material systems called superfluids, which at sufficiently low temperatures tend toward a state of order that is certainly not spatial. This property was first observed for the case of mass 4 liquid helium, which in statistical physics has always played a role analogous to the one played by the hydrogen atom in atomic physics. At the moment we know of three different superfluids, whose basic properties are

Box 2.I

COMPARATIVE SURVEY OF SUPERFLUIDS

Superfluid	*Liquid helium 3*	*Liquid helium 4*	*Super-conducting electrons*
Transition temperature K	0.001–0.003	1.7–2.2	0–23
Physical particles	Atom	Atom	Electron
Statistics	Fermion	Boson	Fermion
Electrical charge	No	No	Yes
Magnetic properties	Yes	No	Yes
Condensed form	Atom pair	Atom	Electron pair
No. of superfluid phases	Three	One	One

compared in Box 2.I. Although this number is rather low, it nonetheless suffices to illustrate the nature of this type of order, in the following terms:

1. The properties of superfluids are unusual because of *the quantum effects operating at the macroscopic scale*. Some extreme examples of these properties are: motion without friction, quantization of the fluid's vorticity, and interference produced by the sum of two currents of matter. This last effect is the macroscopic parallel of wave diffraction of light rays. Of course this occurs only at low temperatures, in the absence of thermal noise.

2. This unusual behavior on the macroscopic scale occurs because a great number of individuals are in the same quantum-mechanical state. Consequently, *the amplitude of the probability for any individual in the system*

becomes authentic space–time wave distributions of matter. There are so many individuals in the same state for reasons required by quantum statistics, because they must always be bosons. Note, however, that a boson can also be made of two Fermi particles with closely correlated spin and momentum, but this correlation should be taken in a statistical sense, because it encompasses all fermions within the coherence distance.

3. In other, equivalent, terms, we may say that *the energy spectrum of superfluids always has one densely occupied level.* In mass 4 helium this level is the system's collective ground state, whereas in helium 3 and in the superconducting electrons it is the single level occupied by the boson pairs, located below the many empty Fermi levels. A population that is so dense and correlated on this single energy level can account for the *united response of the system to external disturbances,* such as the absence of friction displayed by the currents as well as the perfect diamagnetism of the electron sea.

4. Just above absolute zero, disorder can be described in terms of quasi-particles that occupy the upper collective levels. The nature of these quasi-particles varies from case to case. If we raise the temperature even more, the quasi-particle gas becomes denser and denser, until it generates a cooperative transition that breaks superfluidity.

5. In conclusion, let us consider three completely different types of material systems at absolute zero temperature: (a) perfect molecular crystals, (b) Fermi gases, and (c) boson liquids. The third law of thermodynamics says that all three systems must have the same degree of order, despite the different masses and forces embodied in their different macroscopic properties. These three types of system are quite different, but they all display the maximum state of order, and this state is always made up of one unit formed by several correlated parts. These parts are, respectively, the molecules correlated in the ordinary space by the lattice periodicity; the fermions correlated in the momentum space according to the Pauli principle; and the bosons, which for statistical reasons are required to stay in the same state. All of these parts form one unit arising from correlations which at zero temperature achieve the highest possible value. In terms of energy, this unit is reflected in the presence of one single level for crystals and boson liquids, and in an irreducible and sequentially ordered group of levels in Fermi gas. But consideration of the space–time correlations between individuals can offer an alternative description which gives even more information about the system. In the case of boson liquids, and thus of superfluids, these correlations are expressed by one single wave function, shared by all the bosons.

2.9 References

(a) Any text on statistical mechanics can be used for a general treatment of quantum statistics. Particularly, Landau and Lifshitz (26) offer a good explanation of the energy spectrum of boson and Fermi liquids, as well as many of the other theoretical aspects of superfluidity not fully proved in this chapter.

(b) For a simplified presentation of the properties of superfluids, see Wilks (40) or Atkins (2) for helium, and Lynton (29) or Kittel (23) for superconductors.

(c) A very good, simple but not too elementary treatment of all the subjects dealt with in this chapter is found in Goodstein (15), a book which also covers the topics dealt with in the next chapter.

(d) Several more recent developments are dealt with at the elementary level by Mermin and Lee (32) in their article on mass 3 helium, and by Bertman and Guyer (3) in their article on the effect of zero-point energy in solid helium.

(e) Further information, such as the use of ions in liquid helium, can be found in the Varenna summer course, edited by Careri (7) and in the volume on quantum fluids edited by Brewer (5). For many other specific subjects, see the volumes of the series "Progress in Low-Temperature Physics," edited by C. J. Gorter (16). The first volume of this series contains Feynman's article on superfluid helium and the third has the BCS article on superconductivity. Both of these articles give a theoretical presentation that is easy for a graduate student to understand.

CHAPTER THREE

THE SPONTANEOUS ORDERING OF MATTER

3.1 Introduction: Order Change in Phase Transitions

An effective way to study order in matter is to consider the way this order must vary during phase transitions, given that any phase in internal equilibrium must have its own degree of order depending upon its structure. Despite the wide range of order-inducing forces that act upon real systems, phase transitions display quite similar features, and this indicates that there must be a general way to express the establishment of a new order through a phase transition. In this chapter, we shall attempt to identify this general expression using a common fundamental variable, namely the order parameter. This thermodynamic quantity not only expresses the extent of order in the macroscopic system during a transition, but it can also be used to describe the essential features of the new ordered state, thanks to its few internal components. In other words, it drastically simplifies the description problem by identifying the most relevant quantities. Thus by reducing the degrees of freedom of the thermodynamic system, this method proves to be much simpler than the general one provided by statistical mechanics, which identifies the thermodynamic functions in terms of all the levels occupied in the system, and therefore must also include the singularities displayed by these functions at the transition points. However, this statistical-mechanical approach is complex and not direct.

Before we go into the details of the transitions that lead to phase changes, we should point out that it is not always necessary to have a sharp transition to increase the order of the system as the temperature drops. In the first chapter, we saw how every system with several energy levels acquires progressively greater order as the temperature decreases, while no qualitative variations take place in the overall properties of the system. Perhaps the

Box 3.A

THE THREE PHASES OF A SIMPLE MOLECULAR SUBSTANCE

- Let us recall the phase diagram of a normal substance, with particular attention to the critical point *C*.

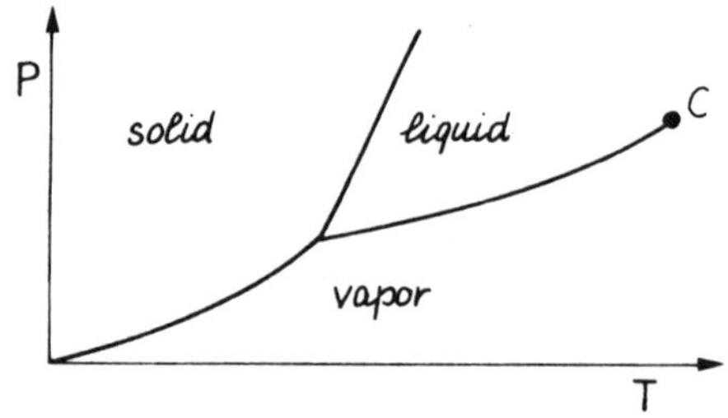

- To measure the disorder in the different phases, we plot the experimental value of the entropy as a function of the temperature (pressure not shown in this plot).

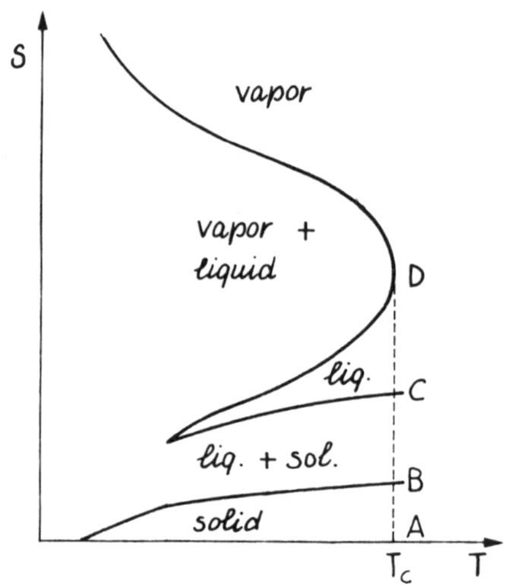

clearest example of this is the paramagnetism of a perfect spin gas, where magnetization is a smooth function of the temperature.

To introduce phase transitions, let us begin with the simplest system, made up of a single component. Generally this system displays at least three phases, which are the liquid, solid, and vapor; two phases can coexist under suitable temperature and pressure conditions, as shown in Box 3.A. The transition that takes place through a coexistence curve is traditionally called "a first-order" one, and it presents a latent heat L and thus an entropy change $\Delta S = L/T$. Only the liquid–vapor transition ends up at a critical point, whose existence was for years a mystery that led to the belief that gases were either "coercible" or "incoercible," depending upon whether they were above or below their critical temperature. We will better understand the meaning of this critical point if we consider the entropy of the three phases, as shown in Box 3.A. For simplicity, isobars have not been recorded in this S, T-diagram, but it is immediately evident that the pressure decreases from the bottom upward, because disorder must increase when the system has a greater and greater volume available to it. The two segments AB and BC along the critical isotherm measure, respectively, the disorder produced in the solid phase by increasing volume, and the disorder due to melting when the two phases are coexistent. The remarkable thing is that the disorder change disappears at point D, and this indicates that two phases are equally ordered at the critical point. This striking thermodynamic property of the critical points will be used in the sections to follow.

Let us say one more word about first-order phase transitions. As we have seen, during these transitions the two phases are in thermodynamic equilibrium with one another, and the entropy change is precisely the measure of the disorder existing between them. Experiments show that the same entropy change takes place in all pure substances during the liquid–vapor transition at atmospheric pressure. This empirical fact, called Trouton's rule, indicates that the liquids on the one hand and vapors on the other must display the same order variations throughout the system because the same disorder is always introduced in the process of evaporation. The only exception to the rule is helium, where the zero-point energy produces more disorder in the liquid phase above the lambda point. The comparison between evaporation entropy and melting entropy is also interesting. Experiments generally show that the former is greater than the latter, and this is explained by saying that the disorder introduced by evaporation is greater than the disorder introduced by melting, because the thermodynamic state of the liquid is closer to that of the solid than to that of the vapor. This must be true, given that melting entropy measures the disorder introduced into

Box 3.B

THERMODYNAMIC FUNCTIONS IN PHASE TRANSITIONS

- Let us record the Gibbs free energy G and the entropy $S = (S(\delta G/\delta T)_p)$ as a function of the temperature at constant pressure.

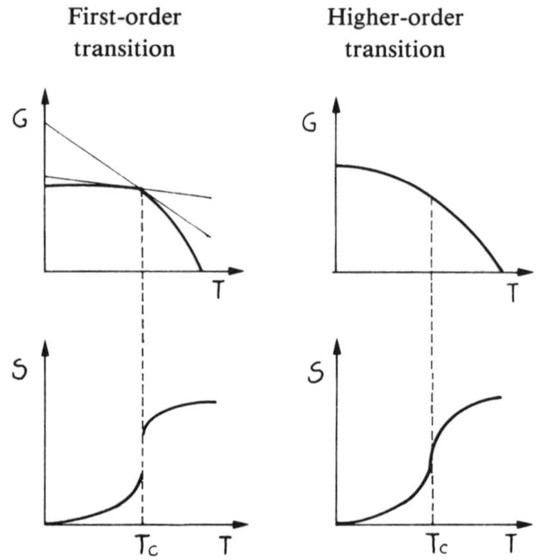

- Note how only the first-order transition displays a jump in entropy and thus can show a latent heat.

- The specific heat $C_p \approx (\delta S/\delta T)_p$ is obviously different in the two cases, and in the higher-order transitions it displays a discontinuity (see Box 2.B for liquid helium and Box 3.F for a binary alloy).

the liquid, but in this way we cannot account for the characteristic order of the solid, namely the long-range order. So we notice that the notion of disorder in a material system offered by entropy may sometimes be rough.

The transition that occurs between liquid and vapor at the critical point is only one case of a wide class of transitions that are sometimes called "higher-order" transitions, but should be called *order–disorder transitions,* or even simply *critical phenomena.* These transitions do not display the sudden state variations that are observed in the first-order phase transitions; furthermore, as we have already seen, the entropy changes quite steadily, as do the other state functions. However, the derivatives of these functions versus temperature display a break or divergence near the critical temperature T_c, and for this reason they are sometimes called lambda-type transitions because of the cusplike form taken by the specific heat. The lambda transition of liquid helium and the para-ferromagnetic transition are two important examples of this type of transition, and in all these cases a most interesting point is the cooperative mechanism by which the new structure is formed.

Before concluding this introductory section, let us further clarify this point by making a comparison between first-order transitions and critical phenomena on a purely thermodynamic basis. Box 3.B shows that a first-order transition results when the two curves representing the free energy in the two phases meet at a given point. Inasmuch as these two curves also have a real meaning far from this meeting point, we should be able to observe such phenomena as hysteresis or metastability, as in the overheating of a solid just above the melting point. On the other hand, nothing of the sort can occur for a transition that takes place at a true critical point, because in this case the free energy is a continuous function of the temperature around T_c. Note again how different are the higher-order derivatives of the free energy versus temperature, a formal difference that originally led scientists to distinguish between first- or second-order transitions, depending upon whether a break was noticed in the first or second derivative. From now on, we shall deal only with the so-called "higher-order" phase transitions, because they are the only ones that display the long-range ordering process as a breaking in the system's symmetry, a notion whose general features will gradually emerge in the following sections.

3.2 Identification of the Order Parameter in Critical Phenomena

To widen our view of critical phenomena, we shall now attempt to identify a parameter η (T) capable of acting as the *basic variable in the description of an order–disorder transition.* To this end, it would be wise to take a simultaneous look at the behavior of two systems we discussed in the first chapter, namely real fluids and ferromagnets. Then we shall extend this concept to such interesting systems as the superfluids we met in the last chapter.

Box 3.C

A UNIFIED PRESENTATION OF CRITICAL PHENOMENA BY THE ORDER PARAMETER

- Consider the phase diagram of two different real systems, each described by its most appropriate variables. In a fluid at $T \gg T_c$, note the limiting equation of the perfect gases, and in the magnet at $T = O$, note also that M does not depend upon H, because there is maximum order.

fluid

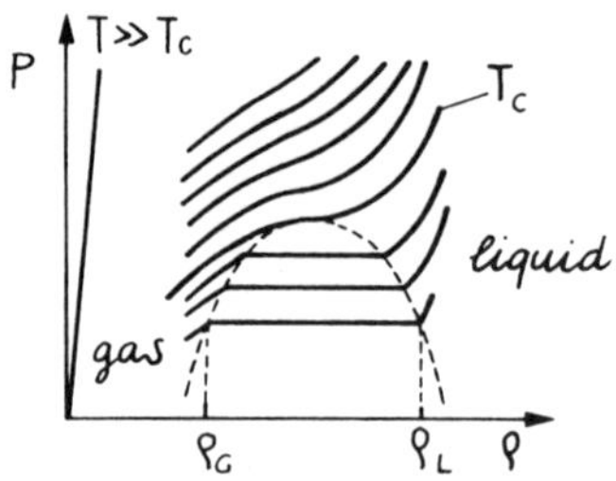

magnet

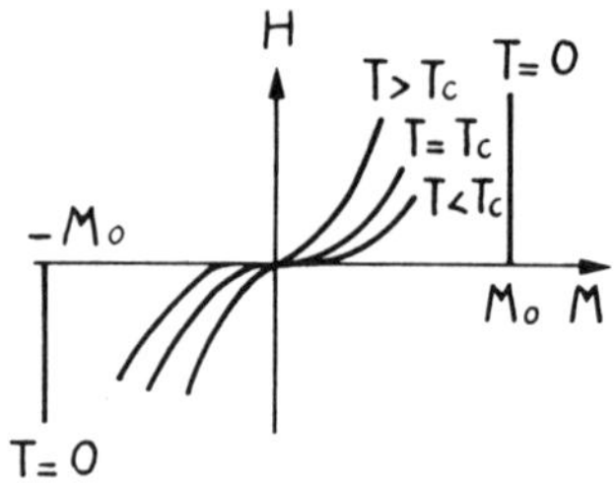

- Let us plot these two critical phenomena as follows:

fluid

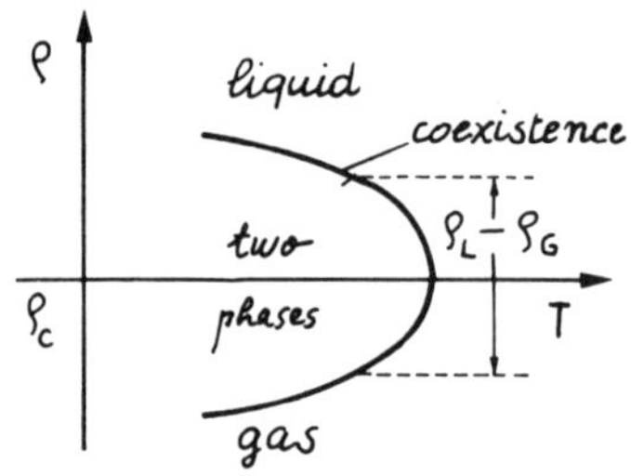

magnet

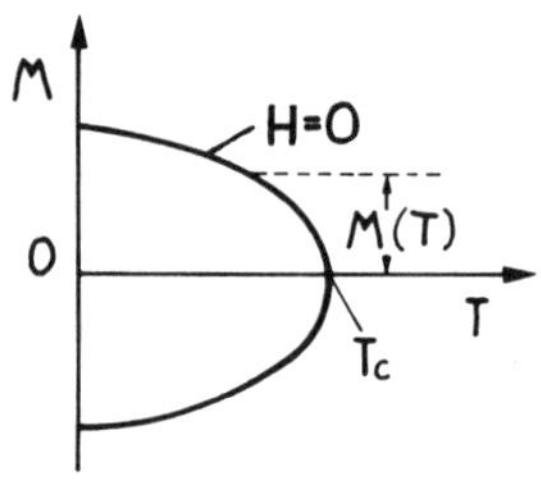

- If the proper definition of the order parameter η is introduced, the behavior of the two systems can be presented together in one plot.

fluid : $\eta = \varrho_L - \varrho_G$ *magnet* : $\eta = M$

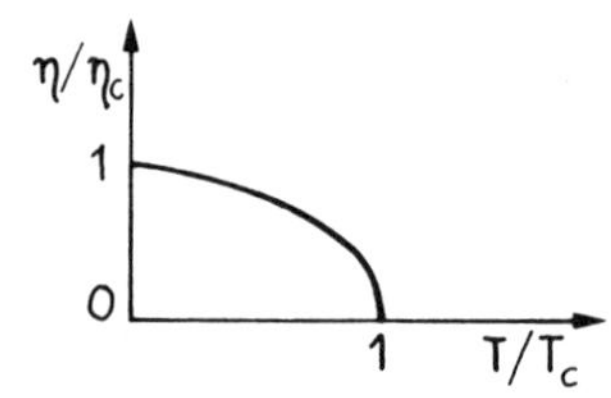

Let's take a look at the figures in Box 3.C. Reading them from the top to the bottom, we see that these two different phase diagrams, each one specific to proper description of its system, end up in a common single plot made by reduced variables. This common plot is possible thanks to the introduction of an order parameter that governs the transition. In Chapter One, we remarked upon the analogy between the critical point in a fluid and the Curie temperature in a ferromagnet. Now we see how this analogy can be taken one step further by combining two intrinsic quantities, namely, density and magnetization. If we introduce a suitably defined order parameter, we can describe the behavior of the two transitions using only one variable in the reduced plot, seen at the bottom of Box 3.C. This fact has already been anticipated in part in the first chapter.

The experimental data of many critical phenomena can be presented in quite a similar description, as long as the order parameter governing the transition is correctly identified. Box 3.D shows some classes of these transi-

Box 3.D

IDENTIFICATION OF THE ORDER PARAMETER IN CRITICAL PHENOMENA

- The order parameter is the fundamental macroscopic variable that makes it possible to formulate a unified description of the order process in very different systems, with critical temperatures T_c that are remarkably different, as shown by the data of this table:

System	*Order parameter*	*Example*	T_c °K
Liquid–gas	Density	H_2O	647.05
Ferromagnets	Magnetization	Fe	1044.00
Antiferromagnets	Magnetization	FeF_2	78.26
Binary mixtures	Concentration	$CCl_4 - C_7F_{14}$	301.78
Ferroelectrics	Polarization	Triglycine sulfate	322.50
Binary alloys	Concentration	Cu-Zn	739.00
Superfluids	Wave function of ground level	He^4	2.18
Superconductors	Wave function of boson pairs	Pb	7.19

tions, and the specific order parameter and one typical example are also shown for each of them. In the ferroelectric transitions, the order parameter is immediately identified because now dielectric polarization takes the place of the magnetic polarization of ferromagnets. This is quite natural, because of the formal equivalence between the electric dipole of the molecules and the magnetic dipole of the paramagnetic ions. The same is true for the antiferromagnetic substances, because these substances display two spin subsystems that accommodate themselves in an antiparallel direction unlike the parallel one taken in the ferromagnets, and the magnetization is always the quantity that properly describes order. All the above cases present some anomalies in electric or magnetic response, and the specific heat displays a striking anomaly near the critical temperature T_c. The case of liquid mixtures or binary solids can also be presented as an extension of real fluid behavior, and we shall return to this type of transition in the following section.

However, in some transitions the order parameter cannot be identified immediately, in the sense that this parameter turns out to be a quantity with several internal components. The case of superfluids is just one example of this. At first glance, we would be tempted to identify the order parameter governing the lambda transition of liquid helium as the density of the superfluid component, introduced in the last chapter through the two-fluid theory and experimentally verified by using a pile of rotating disks. But several considerations then showed us that the characteristic feature of the superfluid phase should be expressed as the wave function of condensed bosons in the ground state, and thus as the two quantities that identify this wave function. For instance, the interference experiments clearly indicate a need for two parameters to characterize the coherent state of bosons. Thus the order parameter of superfluids should really be thought of as a two-component vector. Because the amplitude of the ground-state wave function is not a directly measurable quantity, unlike the density for fluids and the magnetization for ferromagnets, there can be no conjugated external variable to apply to the superfluid, such as pressure to fluids or the magnetic field to ferromagnets. Consequently, superfluids have no measurable quantity analogous to compressibility or susceptibility, and the only measurable critical quantity is the specific heat near the lambda point.

From the viewpoint of the order parameter, the case of superfluids is not an isolated one. As another example, let us consider the transitions that occur in some liquid crystals, which are organic substances made of ellipsoidal molecules that give rise to some intermediate phases between an isotropic liquid and an anisotropic crystal. There are some liquid–crystal transitions in which the molecules change from a structure with parallel layers having no internal regularity, to another structure in which the arrangement resembles a true crystal plane. In this case, too, the order parameter governing the transition must have two components to characterize the spatial order of the molecules with a wavelike periodic space variable.

This analysis shows that the order parameter is a specific feature of the material structure and as such cannot be artificially introduced *ad hoc* from without. *Thus the order parameter appears to be a multicomponent quantity,* like a vector in a suitable space, and the information it can express about order is more than can be supplied by mere negative entropy change. This is because entropy measures the volume of the phase space in terms of the number of quantum states or of the corresponding occupied energy levels, and the information supplied by the value of an energy level differs from that offered by the corresponding space–time wave function. Once again, the case of superfluids can clarify this point, because the information given by the wave function that characterizes the boson state is not the same as the information offered by the presence of a single occupied collective level. Below, we shall try to describe the spontaneous ordering of matter by making systematic use of the order parameter.

3.3 The Mean-Field Theory of Cooperative Order

As we mentioned in the first chapter, van der Waals (for fluids) and Weiss (for ferromagnets) tried to offer an idea of the critical behavior by introducing a mean internal field of a suitable intensity. Because these phenomena occur in conditions of thermodynamic equilibrium, the correct value of this mean field must be obtained by minimizing the free energy of the system versus the parameters that describe this internal field. Although this was not the original procedure used by van der Waals and Weiss, this thermodynamic criterion is quite general, and that makes it the most useful one, because it can be applied to several different systems. In Chapter One, we introduced the meaning and the limits of the mean-field notion when dealing with Einstein's solid. Now let's try to describe a typical critical transition from the viewpoint of the mean-field theory, allowing the order parameter to play its detectable role which will make it possible to generalize the conclusions we shall reach below for a particular case. The one we shall consider is the order–disorder transition of a binary alloy, because this system's order parameter can be so easily identified.

The order–disorder transitions in binary alloys are characterized by a lambda-type anomaly displayed by the specific heat, and by the simultaneous appearance of a *superstructure* below the critical temperature T_c. The term superstructure is used here to mean a long-range order that can be detected by x-rays and is destroyed above T_c. One classic example of this is a brass alloy composed of equal parts of copper and zinc, which crystallizes in the body-centered cubic system. In this case, the superstructure can be described as the interpenetration of two cubic subsystems, one made solely

Box 3.E

ORDER–DISORDER TRANSITIONS IN ALLOYS

- In a brass alloy with equal concentrations of copper and zinc, x-rays show that as the temperature rises a crystalline structure with ordered atoms undergoes a transition to another structure that differs only in the atomic positions.

  In two dimensions

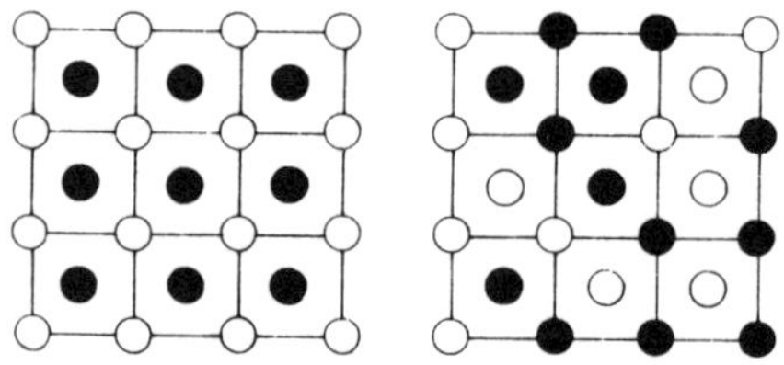

- To describe this transition, we may define the order parameter as the average quantity $\eta = [N_1(1) - N_1(2)]/[N_1(1) + N_1(2)]$, where $N_1(1)$ = number of the type-1 atoms in place 1, and so on.

  Note that when all the atoms are in the right place $\eta = 1$, whereas when only half of them are in the right place $\eta = 0$. These are extreme cases of long-range order and disorder, respectively.

- In this mean-field approximation, the calculation of the optimal thermodynamic quantities against η shows the plot of the order parameter η and of the specific heat C as shown below:

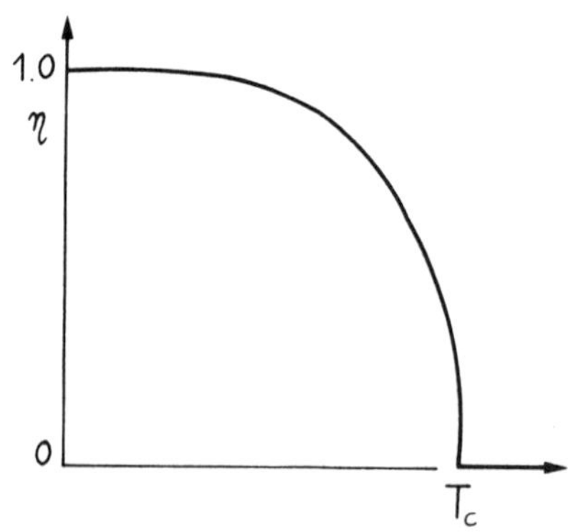

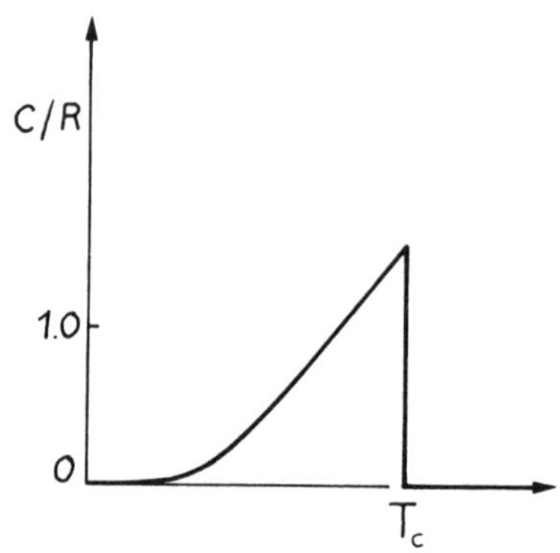

of copper atoms and the other only of zinc atoms, so that in the most ordered state, every atom of one type is surrounded by eight atoms of the other type. As the temperature goes up, a greater and greater number of atoms accommodate themselves in the wrong crystalline sites, until a completely disordered state is obtained when every subsystem contains an equal number of both type atoms. Consequently, in this disordered state one can no longer detect the contribution of certain pairs of lattice planes by x-ray reflection, and the superstructure disappears even while the geometry of the body-centered lattice remains intact.

In the case of this binary alloy, the order parameter can be immediately identified as a function of the mean number of type-1 atoms that occupy the right place $N_1(1)$ and the wrong place $N_1(2)$, as defined in Box 3.E. Next, it is a simple combinatory computation to write the internal energy E as the sum of the energies of all the pairs, once the interaction energies between equal and unequal neighboring atom pairs is given, and it also is easy to write the entropy S of the system as one of a binary mixture, again as an explicit function of the order parameter. At this point, the free energy $F = E - TS$ must be minimized versus this order parameter. We thus obtain its optimal value, which characterizes equilibrium, and which itself proves to be a function of the temperature and the given energy constants. Unfortunately, this calculation does not yield a simple analytical function, but this function can be plotted and it gives an order-parameter curve like the one seen at the bottom of Box 3.C and stressed in Sec. 3.2 as the typical one for a critical phenomenon.

Once the optimal value of the free energy has been established, the usual formulas of thermodynamics can be used to identify all the other thermodynamic functions desired. One of particular interest is the specific heat $C(\mathrm{T})$, which shows a cusp centered around T_c, as shown in Box 3.E. At first, this fine result gives the impression that such a simple theory is actually capable of describing the cooperative effect that leads to the order–disorder transition. Furthermore, this result can be immediately extended to other critical phenomena, because the theory is so general, using as it does only a thermodynamic stability criterion, and because of the simple and evident role played by the order parameter, that of an average quantity of clear significance.

Incidentally, we should note that the transition of this binary alloy is quite useful to introduce the *symmetry-breaking* notion, to which we shall return in its more general form. Indeed, it is clear that the disordered phase of this alloy displays a greater symmetry than the ordered phase, because in the former more translation operations are possible among the lattice planes. The superstructure is the most ordered phase, but it is less symmetrical because the only possible translations are the ones that do not alter the distance between the lattice planes containing equal atoms. Thus we can begin to visualize the ordering process as a *reduction in the number of operations that can be performed while keeping the system invariant,* thanks to some new constraints leading to the formation of new relationships between the relevant quantities in play.

Landau made the next approximation of this mean-field theory, using a series expansion of the free energy as a function of the order parameter, by adding one more term to represent the space gradient of the order parameter. At any rate, it should be clear that even this improved mean-field theory assumes that the mean value of the order parameter is the most prob-

Box 3.F

SOME CRITICAL EXPONENTS

- Many physical quantities display an unusual behavior at the critical point and can be described using a suitable exponent near T_c, such as:

  Specific heat $\approx (T_c - T)^{-\alpha}$

  Magnetization of ferromagnets
  Difference in density between liquid and vapor } $\approx (T_c - T)^{\beta}$

  Magnetic susceptibility
  Fluid compressibility } $\approx (T_c - T)^{-\gamma}$

- It is remarkable that the experimental values of these exponents turn out to be so close, even among very different systems, but are far from the value predicted by the mean-field theory.

Material	T_c°K	α	β	γ
Fe	1044	0.120	0.340	1.33
Ni	631	0.100	0.330	1.32
CO_2	304	0.125	0.340	1.20
Xe	289	0.080	0.330	1.20
He^3	3.31		0.361	1.15
He^4	5.18	0.127	0.355	1.17
Mean-field theory		0.000	0.500	1.00

able value and that it is a macroscopic quantity unrelated to microscopic scale quantities. A similar assumption is made in classic hydrodynamics where the waves of a fluid are not related to the molecular processes but obey their own laws, for which the molecular properties are included in properly averaged parameters such as the density, the elastic constants, and so on. In this case, too, the most correct method would be to consider the spontaneous fluctuations at every scale of the order parameter and then average them over all the microscopic states of the system. As we shall see later, this is the essence of the more modern theoretical, so-called "renormalization group," approach.

Even in the form proposed by Landau, the classic mean-field theory reached its limits when experiments were carefully performed very close to the critical temperature. As we have already seen, many physical properties diverge near T_c, so that it is better to express the experimental results in terms of powers of the displacement from the critical point, thus defining some suitable critical exponents. The mean-field theory gives the value of 0.5 for the order parameter, but accurate experiments have given much lower values for this exponent, as can be seen in Box 3.F. The same is true for the specific heat, which in the frame of this theory displays a discontinuity that cannot be expressed by means of a power series, as well as for other typical response functions such as the isothermal compressibility of fluids and the isothermal magnetic susceptibility of ferromagnets. If we examine the data shown in Box 3.F, we shall see how close to each other the experimental values of all these quantities still are, despite the great diversity of systems considered, so we should conclude that the lack of agreement between experiment and theory must include some general and systematic features.

As we have said, the weak point of this mean-field theory is that it completely ignores the order-parameter fluctuations. Thus, in the next section, we shall take a look at other experimental properties near the critical point, in order to gain some insight into the possible presence of space–time fluctuations in the order parameter.

3.4 Space Fluctuations of the Order Parameter

One remarkable occurrence that invariably takes place in fluids at the critical point is the appearance of a strong *opalescence,* namely, a sudden increase in the light scattered by the fluid. The same thing happens near the demixing point in transparent binary-liquid mixtures, systems that are similar in many ways to the brass alloy considered in the preceding section. At the turn of the century, Smoluchowski and Einstein tried to explain this phenomenon by attributing it to the density fluctuations occurring in the body. It was immediately thought that in order to generate the milky-white color of scattered light, these fluctuations ought to encompass the entire wavelength spectrum, and that they should also be of quite high amplitude, to justify the extent of the effect observed. But the correct theory was formulated shortly thereafter, by Zernike and Ornstein, who introduced the space correlation between density fluctuations, a quantity that they thought should extend to long distances near the critical point.

The quantitative interpretation of critical light scattering is based on two theories in physics, namely, the statistical theory of fluctuations and the electromagnetic theory of scattering, both of which require extensive analytical development. For this reason, we shall limit ourselves to examining this phenomenon from an oversimplified viewpoint, so as to understand how the scattering of a wave traveling through a material system can supply useful information about the system's order. Let us begin with the well-known case of the diffraction of x-rays by a crystal, which takes place because every atom in the lattice behaves like a scattering center with respect to the incoming electromagnetic wave. Now let us reformulate this by saying that the light beam is scattered only because the microscopic dielectric constant of the body, which is made up of atoms located at the lattice nodes, varies from point to point. As is well known, only when the crystal is truly perfect does the x-ray beam rebuild itself in the direction of propagation, thanks to the contribution of all the scattering centers, allowance being made for the Bragg scattering which occurs only for some specific angles of the impinging beam. Now if we generalize this picture to a system that is disordered due to the presence of impurities or empty bubbles, we can immediately see that glass becomes opalescent if the dimensions of these bubbles are close to the wavelength components of the light beam. A light beam is a coherent phenomenon, and thus it reacts to the lack of order in the body it is crossing, by lowering its forward path and spreading out. In this way, the intensity of the scattered light can give information about the number of impurities and, more generally, about the extent of disorder, whereas the relationship between the observed effect and the wavelength employed also gives information about the space dimensions over which the disorder is spread. In principle, this is an ideal way to study the density fluctuations in a body, because this kind of measurement causes minimal disturbance to the body being examined. In the case of a crystal at low temperatures, these density fluctuations are the phonons we have already met as local elements of disorder in a highly ordered aggregate. But although wave-like propagation laws are valid for phonons, this is certainly not true for fluid regions at the critical point, when the propagation of sound goes to zero because the fluid's compressibility becomes infinite. Thus the phenomenon of critical opalescence points to the presence of an entire range of disordered regions spread at random throughout space and time. Now, if we turn this argument around, we realize that an extended order must also appear in the critical phenomena. To be somewhat more accurate, we may say that this order appears as a correlation among otherwise random fluctuations of such microscopic elements as the spins or the atoms; furthermore, we see that this correlation extends over ranges that are wider and times that are longer than those typical of the microscopic elements. This is an essential point, and we shall return to it several times in this chapter in order to clarify it.

The previous picture is perfectly general, and it holds even when the incoming wave is associated to a material particle, as in the case of a beam of neutrons of the same velocity. Thus this technique of critical scattering can be modified and adapted to the conditions that best fit the type of matter and disorder to be studied. So x-rays are used for the binary-alloy transitions considered above, whereas low-frequency, electromagnetic waves are used for structural transitions in ferroelectric matter, and light waves are used for transparent liquid mixtures. But what is more, slow neutrons are used in a wide variety of structural transitions in solids and in magnetic systems. In the ferromagnetic transitions, it is found that the neutron scattering increases as the Curie temperature is reached, and this holds true no matter what the energy of the neutron. Given the formal equivalence between light frequency and neutron energy, this phenomenon is the exact analog of the opalescence found in fluids at the critical point. But whereas in fluids the frequency distribution of the scattered light is sensitive to some other dynamic effects, in ferromagnetic substances we can study the forward scattering fairly easily in the long-wavelength region or, to be more exact, when the propagation vector tends toward zero without an energy change.

The experimental interest for the critical scattering at long wavelengths stems from the fact that the mean-field theory, which also includes the space gradient of the order parameter, states that the dimensions within which the order parameter is almost constant must increase dramatically near the critical point. We can get a better idea of this if we think of a ferromagnet and call *coherence distance* the linear dimension within which the spins temporarily remain parallel to one another, to give rise to a local magnetization. In this case, the mean-field theory predicts a divergence with an 0.50 exponent for the coherence distance, whereas the long-wavelength neutron technique for the ferroelectric transition of iron detects a value of 0.65. This discrepancy between the mean-field theory and experience is very important, and we shall return to it in the next section, when we shall add a time character to these wide-scale space inhomogeneities.

Before ending this section, we should emphasize that when we say that the coherence distance is much greater than the microscopic dimensions of the system, this must imply that the critical phenomena must display a long-range order process. As we have already mentioned, such a coherence distance can be expressed more correctly as the distance over which are extended the correlations between the local fluctuations of the microscopic bodies that make up the system. In other words, the spins cannot undergo independent fluctuations within one domain, and thus the dimensions of the domain must coincide with this coherence distance.

3.5 Dynamics of the Order–Disorder Transitions

In Sec. 3.4, we discussed the behavior of some static quantities. But when a system approaches the critical point, important anomalies begin to appear in several of its dynamic properties, such as the diffusion and viscosity coefficients, the frequency distribution in light or neutron scattering and, more generally speaking, the time the system takes to relax after every type of external disturbance. Unlike the static phenomena, for which every property can be traced back to the statistical distribution at equilibrium, the dynamic properties require knowledge of the equations of motion governing the temporal evolution of the various parts of the system. Thus we must expect many interesting dynamic phenomena, but this means they will be harder to interpret and generalize.

On the other hand, if we are interested only in the time behavior of the order parameter's wide-scale fluctuations, it is not hard to identify a general experimental occurrence called *critical slowing down,* which means that the order parameter must vary at a slower and slower rate as the critical point is approached. We can establish this idea more clearly by returning to the ferromagnetic transition, where we know there are domains inside which the spins run parallel. It stands to reason that the wide dimensions of these spin domains make it take much longer for thermal agitation to rotate this whole spin group randomly. In more suitable and general terms, it may be said that *long-wavelength modes are very slow near the critical point.* To explain how all this occurs, a microscopic theory should consider the interactions between spins and the laws of dynamics governing the spin motion, a difficult topic that does not directly interest us here.

Let us pause a moment to visualize the behavior of matter at the critical point; we shall have a clearer outlook if we consider a fluid and a magnetic system at the same time. We have seen that the order parameter is zero near the critical point, when the coherence distance extends to a large region and the correlation function drops off very slowly over time. Thus in the case of the fluid we must think in terms of large and denser areas separated by other large but less dense areas that slowly form and break up so as to keep the mean value of density constant over the whole volume occupied by the fluid. Of course, we are dealing with a whole range of regions including those even smaller than the coherence length, but the larger regions are the most persistent ones in time. The same thing occurs in the magnetic system, where domains are formed by internally parallel spins, but these domains are randomly arranged so that the resulting magnetization of the whole system is zero, whereas there do exist local fluctuations of the magnetization caused by the presence of domains. Once again, there is a wide range of sizes for these domains, but the larger ones have the longer temporal decay constants. Thus, we may summarize this picture by saying that both the instantaneous value of the order parameter averaged over the total region available to the system, and its local value averaged over a rather long time, are zero at the critical point, whereas on lower space and time scales these same quantities are certainly not zero.

This reasoning shows that we must identify a few *slow modes* whose relaxation time goes to infinity at the critical point, if we are to formulate a phenomenological description of dynamics. Some of these modes interact in a wavelength scale comparable to the coherence length, a quantity, however, that has been seen to diverge at the critical point. Some of these nonlinear interactions are dissipative and are responsible for the fact that several irreversible transport processes approach zero at T_c, but there are other interactions among these slow modes that are entirely reversible and are responsible for the divergence of other kinetic processes at T_c. Unfortunately, the relevant specific dynamic laws make this behavior far less general than the one we observed for the static properties.

Today we are certain of the existence of these slow modes in many material systems, thanks particularly to the neutron-scattering technique. For many structural transitions that occur in the solid state, these very slow modes give the key to the order–disorder transition mechanism and also allow a direct identification of the order parameter. In these cases, the critical transition can be observed as the collapse of a crystalline structure because the frequency of a crystal vibrational mode goes to zero when the temperature approaches T_c. Indeed, at T_c the amplitude of some slow lattice vibrations can become excessively large for energetic reasons, and may produce some permanent atomic displacements that generate the new crystalline structure because the former one has become unstable. Box 3.G shows the atomic displacements that lead to the ordered phase below T_c in some of these transitions. In these cases the measured displacement amplitude is the very order parameter, and it displays the typical behavior already seen. Often it happens that another vibrational mode is also excited by the slow mode because of coupling between modes, and thus it contributes to the collapse of the crystal structure as well. For this reason, these structural transitions lead to such catastrophic and abrupt changes that they sometimes look like first-order transitions. It is also important that the existence of a particular dynamic feature, as shown by these slow modes, can be related to a general property observed in all the order–disorder transitions, namely, the breaking of some of the translational space symmetries. We shall return to this important fact in the next section.

Box 3.G

ATOMIC DISPLACEMENTS IN STRUCTURAL TRANSITIONS

In some phase transitions where a transformation of the crystal lattice takes place, it is possible to directly identify lattice slow modes, which become wider at the critical point and cause the structure to collapse above some value of the atomic displacements. In these cases, the amplitude of these displacements is the order-parameter value appropriate to the transition and can be measured by neutron scattering. For instance,

Acoustic mode

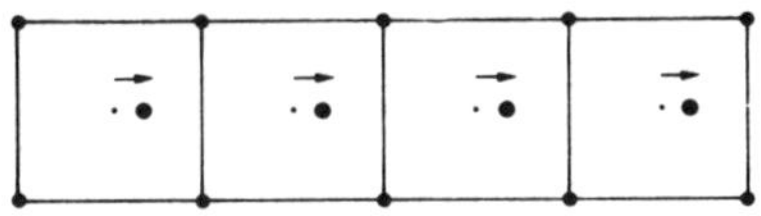

Optical mode

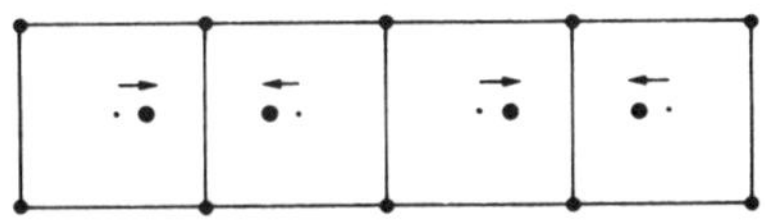

Librational mode

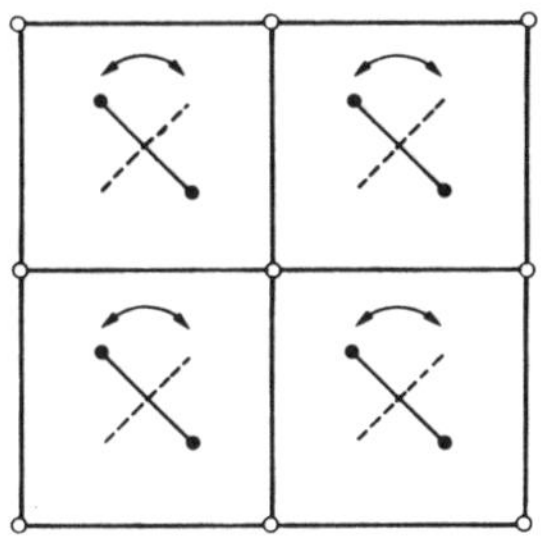

3.6 Order and Symmetry Breaking

"Symmetry breaking" is used to mean the formation of a new situation in which the system's state no longer possesses all the symmetry characters that its interactions would allow it to have, so that it can only take a more *restricted number of configurations.* For instance, when the temperature is lowered, a cubic system may undergo a transition to hexagonal, and this is called a symmetry breaking. But there will be no transition from hexagonal to cubic because the latter has the greater number of symmetry elements.

The lattice-symmetry changes in phase transitions satisfy several general theorems which prove that the symmetry characters in the two differently ordered systems must always be closely related. This is certainly not true in the first-order transitions, so that symmetry breaking is the most suitable way to characterize phase transitions, rather than the traditional thermodynamic criterion mentioned at the beginning of this chapter.

It can also be proved that when the symmetry group being broken is a continuous one—characterized by invariance towards operations such as rotation or translation with a continuous value of angular or linear displacement—then the lowest symmetry phase must display the so-called Goldstone excitation whose frequency must go to zero at the greatest wavelengths. For instance, in the case of a multidimensional spin system, above T_c every spin may rotate at any angle with respect to its neighbors, but this is not possible below T_c, and the group of excitations that are consequently generated are the spin waves themselves. Indeed, it is clear that the frequency of these waves must go to zero at T_c because at this temperature it no longer costs any energy to rotate one part of the system with respect to another. If we extrapolate this scheme, we may imagine that when a liquid freezes, a solid forms with a phonon branch, because of the breaking of a translational symmetry that is complete in the liquid and then reduced to the smaller number of translations allowed in a crystal. More generally speaking, we may say that this justifies the appearance of the slow modes in the ordered phase, as mentioned in the last section. Inversely, when symmetry breaking affects a discontinuous group, as in the Ising model where the magnetization can set up or down but never in the intermediary positions, there is no longer formation of a continuous spectrum of modes. This reasoning thus shows how *it is possible to connect the order parameter to the symmetry properties of the system through its dynamics.*

Recently, it has become more and more apparent that symmetry breaking is a specific phenomenon of the material systems in the condensed state, and that it has several intrinsic characteristics and is independent from boundaries and from external disturbances. The characteristics of this phenomenon can be summarized in the existence of some general theorems concerning the symmetry changes allowed in phase transitions and in the dynamic consequences that lead to *soft modes,* as pointed out in this section, but also in some other features worth recalling here. First, the less symmetrical state in superfluids displays those long-range effects that cause the more ordered system to react with rigidity to every external disturbance. We have seen examples of this rigidity in the infinite diamagnetism of the superconductors, as well as in such topological defects as the vortical lines in superfluids, that allow this rigidity to relax at some intermediate scales. Furthermore, real material systems tend not to set exactly at the symmetry-breaking point but at both sides of it; namely, they tend to fall into the two borderline conditions. And finally, the more ordered situations are reached without the series of instabilities such as the hydrodynamic instabilities generated at high fluxes that will be seen to characterize the ordering process in open systems. We must stress these points to avoid the improper way the term *symmetry breaking* is sometimes used, when only one of the above aspects is actually observed.

3.7 Dimensions of the Order Parameter

In the preceding sections we saw how the use of the order parameter makes it possible to have a unified description of all critical phenomena. Further analysis of the divergencies observed at T_c will now enable us to see how some universal classes can be formed, within which different material systems display identical critical exponents, and how these classes are characterized by only two quantities. One of these quantities is the dimensionality d of the physical system, and the other is the dimensionality n of the order parameter we have already met. This important general fact is illustrated in Box 3.H, which we shall now discuss. In this box, it is striking to see *the importance of d and n, but also the irrelevance of every microscopic detail in the formation of these universal classes.*

A simple explanation can be offered to show how the dimensionality of the physical system must be relevant for a cooperative transition. Suppose we have a group of individuals among which a signal can be transmitted only between first neighbors. In this case, there is a good probability that the signal will be transmitted over a large distance if the individuals are arranged in two dimensions, and it would improve if they were in three dimensions, but it would decline if they were in only one dimension, where even one error would fatally compromise correct transmission of the signal. The same is true for an effect that must be coherently propagated in a physical system, and thus for critical phenomena where a long-range order must be formed. Indeed, the theoretical study of cooperative Ising-like models, with interactions only between first neighbors, clearly shows that the critical exponents depend upon the space dimensions of the system in question.

The most interesting fact is the importance of the dimensionality n of the order parameter. This number is connected to the degree of symmetry or isotropy of the order parameter itself. In fact, when $n = 1$, the order parameter is a scalar quantity, as in fluids where it is the density change, or in binary mixtures where it is the concentration. When $n = 2$, the order parameter is a two-component vectorial quantity, as for the superfluids and some liquid crystals, where it shows a wavelike behavior described by the amplitude and phase of the wave. The magnetic systems fall into the $n = 1$ category if magnetization occurs along one axis only, whereas they must be classified $n = 2$ or $n = 3$ if the spin vector can be, respectively, set in one

Box 3.H

THE UNIVERSAL CLASSES OF CRITICAL PHENOMENA

- Let us define *d* as the dimensionality of the physical system, and *n* as the dimensionality of the order parameter.
- The universality hypothesis states that when *d* and *n* are the same, all the systems must behave the same way, as seen in the table below:

Universality class		*System*	*Order parameter*
$d = 2$	$n = 1$	Absorbed films	Surface density
$d = 2$	$n = 2$	Superfluid He^4 film	Superfluid wave function
$d = 3$	$n = 1$	Uniaxial ferromagnets	Magnetization
$d = 3$	$n = 1$	Fluids	Density difference
$d = 3$	$n = 1$	Mixtures, alloys	Concentration difference
$d = 3$	$n = 2$	Planar ferromagnets	Magnetization
$d = 3$	$n = 2$	Superfluids	Superfluid wave function
$d = 3$	$n = 3$	Isotropic ferromagnets	Magnetization

plane or in ordinary space. Naturally, this order-parameter symmetry is not directly related to the classes of geometrical symmetry of the material system, although both are intrinsic properties of the system. To further emphasize the symmetry character of the order parameter, we may say that $n = 1$ corresponds to linear symmetry, $n = 2$ to cylindrical symmetry, and $n = 3$ corresponds to spherical symmetry. For this reason, n is also called the isotropy index of the order parameter.

Now let us go back to Box 3.H to consider the so-called *universality hypothesis,* which states that different critical phenomena must behave with the same critical exponents when n and d are the same. According to this hypothesis, the (d,n)-plane must be considered analogous to the Mendeléev table of chemical elements, in the sense that it must be possible to predict the properties of a critical system with some given d and n values, from the knowledge of the properties of other systems already experimentally known. For instance, near T_c a uniaxial ferromagnet must behave like a brass alloy, a planar ferromagnet like a superfluid, and so on. Just as the noble gases are a constant reference point in the periodic system of the elements, even though they are not chemically interesting, the same is true for the infinite and the zero cases of n and d. Finally, even within the (d,n)-plane we may find some correlations between the values of critical exponents belonging to different lines or columns.

The theoretical justification of the (d,n)-plane properties was only recently formulated by Wilson and Fisher, using a method called the renormalization group. This technique makes it possible to deal with problems in which many degrees of length or energy come into play at the same time, such as the critical phenomena where all the order-parameter fluctuations from atomic distances to the coherence length are of equal importance. By averaging the fluctuations first on a small, well-defined scale, such as between 2 and 4 Å, and keeping proper account of this result while calculating the next average at the next larger scale, and thus proceeding to include all the scale lengths, this technique can amplify or eliminate several quantities and assess the essential aspects of the phenomenon. Thus in critical phenomena the small-scale fluctuations are eliminated, and this justifies the hypothesis of universality that in fact overlooks such microscopic details as the lattice parameter or the molecular dimensions. Inversely, only the wide-scale fluctuations are amplified, and thus this fact characterizes the critical point for the appearance of correlations extending to an infinitely long range. So this justifies the so-called *scaling hypothesis,* which states that there is a correlation length, which we have seen to be the mean distance within which the fluctuations are correlated, and that this is the only relevant length in the critical phenomena. Thus because the correlation length diverges at the critical point, this very divergence must also affect all other quantities whose physical dimensions depend upon length.

In the renormalization-group calculations, d and n appear in equations where they can take on any continuous value and not only the integer and positive ones that make physical sense. Thus because it was known that the mean-field theory gave exact results for $d \geq 4$, it was possible to extend this theory to the cases of physical relevance with $d = 1,2,3$, by expanding the critical exponents as a function of d near the $d = 4$ value, and thereby obtaining figures in extremely close agreement with the experimental ones. We should not be surprised if the critical exponents merge into the mean-field theory predictions when the dimensions of the system are increased, because it is obvious that the number of neighbors increases along with the space dimensions of the system. But why $d = 4$ is such a sharp limit, above which the mean-field theory gives exact results, is still a mystery. The contour lines of the order-parameter exponent calculated using this method are shown in the (d,n)-plane in Box 3.I.

Box 3.I

THEORETICAL VALUE OF THE ORDER-PARAMETER EXPONENT

- Near the critical point, write the order parameter

$$\eta(T) = (T_c - T)^\beta$$

and plot the constant β curves on the (d,n)-plane calculated according to the renormalization-group theory.

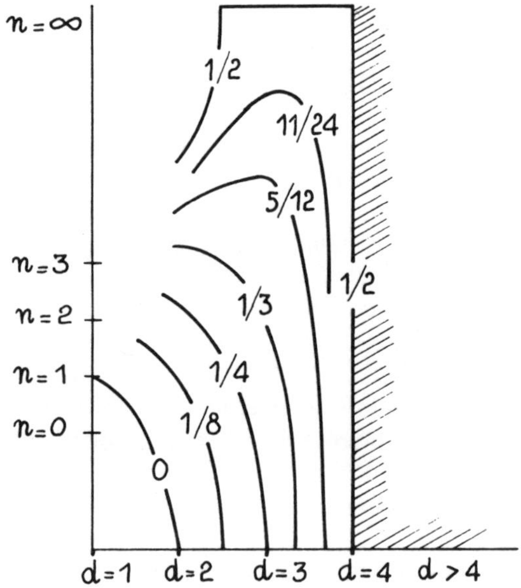

(Simplified presentation adapted from the article by Wilson [42])

- Note that in the shaded area, β is always 1/2, which is the value predicted by the mean-field theory.

Thus, we can be absolutely certain that *for all physical systems in ordinary space we must correct the mean-field theory by introducing the order-parameter fluctuations.* This is the conclusion of the line of thought we started in Chapter Three with the introduction of the wave function as the order parameter in superfluids, and then generalized in this chaper to include all material systems displaying critical phenomena when an order parameter can be identified. The spontaneous fluctuations of some specific properties can always be correlated in the different states of matter, but what characterizes all critical phenomena is the fact that the correlation distance of the order parameter tends to infinity at the critical temperature. In the next chapter we shall try to extend this line of thought to the critical phenomena far from thermodynamic equilibrium.

3.8 Summary and Conclusions

In this chapter we have seen how order and disorder can coexist in equilibrium at a given temperature. To this end, we have carefully considered the behavior of critical phenomena, because they make it possible to identify the order change as the only element of diversity in the same system. Thus we have concluded that:

1. A change of order induces a qualitative change in the major physical properties of the material system. Many of these properties diverge near the critical temperature T_c, and these divergencies can be described by suitable critical exponents. The entropy varies continuously going through T_c, whereas the specific heat change is discontinuous. The order–disorder transition can be described as a spontaneous cooperative effect caused by interaction among the individuals of the system. In the mean-field theory, this interaction is averaged out, but the result is that the theoretically predicted critical coefficients are not in agreement with the experimental values.

2. There is a macroscopic quantity, called the order parameter, which works as the fundamental variable that makes it possible to formulate a joint description of critical phenomena that are apparently quite different. For real systems, it is necessary also to consider the order-parameter fluctuations in all length ranges. Specifically, the coherence length, defined as the greatest distance within which the fluctuations of the order parameter are still correlated, goes to infinity at the critical temperature.

3. The order parameter displays a small number n of components, which may be seen as isotropy or symmetry indices determined by the nature of the

system itself. For instance, in the case of a superfluid, $n = 2$, because the boson-ordered phase is described by a wavelike function with two parameters. The dimensionality n of the order parameter and the dimensionality d of the system's physical space are the only quantities that allow the formation of universal classes for all the critical phenomena on the (d,n)-plane, with no relevance for the microscopic details of the systems.

4. The ordering process that takes place in critical phenomena can be visualized as a symmetry breaking, in the sense of a restriction in the number of invariant transformations and thus as the imposition of more relationships that characterize the state of greater order. Because the new state of lower symmetry makes the system more rigid, this calls for a series of characteristic effects, such as the appearance of soft modes below the critical temperature, and so forth.

5. Study of phase transitions has shown that we must consider the correlation between the spontaneous fluctuations of the system. This essential concept combines both notions of order and disorder in matter at equilibrium. We shall also use this concept later on to describe some phenomena far from equilibrium.

3.9 References

(a) Unfortunately, there is no truly elementary treatment of the critical phenomena. The best introductory text is Goodstein (15), based upon the classic work by Landau and Lifshitz (26).
(b) For more details on critical phenomena from both the theoretical and experimental viewpoints, consult several articles in the series edited by Domb and Green (11), the book by White and Geballe (39), and the book edited by Burgess (6).
(c) For the most recent theoretical developments concerning the renormalization group, see the introductory article by Wilson (42) and the fine book by Ma (30) for a further study. For more elementary topics, see the article by Fisher (13) and the book by Stanley (38). The Varenna course edited by Green (18) also contains several articles that are still valid.
(d) The theory of thermodynamic fluctuations is handled in every text of statistical physics. Particularly suitable are the books by Goodstein (15) and of course Landau and Lifshitz (26). Some applications of this theory are covered by Kittel (22) and McDonald (31).

CHAPTER FOUR

INDUCED ORDER IN OPEN PHYSICAL SYSTEMS

4.1 Introduction: The Physics of an Open System

One fundamental difference between closed systems in internal equilibrium and systems open to fluxes of matter or energy lies in their behavior towards time. In equilibrium, every flux in one direction is compensated by another in the exact opposite direction, every production is cancelled out by a corresponding loss, and every event is likewise wiped out by an opposite event, so that the system remains in an invariant state towards time reversal. This symmetry is suddenly broken if the fluxes entering the system can move it to a state that is sufficiently far from equilibrium, a new state that seems to be characterized by more order. In this chapter we shall see how to describe this ordering process in some physical systems far from thermodynamic equilibrium, and particularly, how these systems can remain in this stationary state of order while being subjected to external fluxes. By now, many different systems have been observed in this ordered state induced by external fluxes, and recent studies have revealed that they have surprisingly similar behavior, which makes a certain generalization possible. Our interest in these simple open systems stems from the fact that some of their induced disorder–order transitions can themselves be used as a model for much more complicated situations occurring in such interesting natural processes as the biological ones.

Let us begin with the notion of *open system,* one into which fluxes of energy and matter can come and go, thanks to the ability to exchange with the environment, so that the system can be displaced from thermodynamic equilibrium and kept away. For instance, radiating energy may enter the system while thermal energy leaves it, so that a stationary state is maintained. Or a few free energy-rich chemical products may enter while many other, far poorer, ones may leave. These examples are an immediate indication of analogies with what takes place in the biological world. What is more, these fluxes may be varied from outside so as to control the system's evolution towards situations that are farther and farther away from equilibrium. In all open physical systems studied so far, at a critical value of the external flux a transition takes place toward new structures which even qualitatively display properties that differ from the original ones because of the presence of greater order. This is what is meant today by phase transition or critical instability in a system far from equilibrium. Of course, there is no contradiction with all that we saw in the last chapter regarding closed systems in thermodynamic equilibrium, because now the external fluxes invariably raise the free energy level in the system. Indeed, it is precisely the displacement of free energy with reference to its isothermal value of equilibrium that gives the correct measurement of the displacement away from thermodynamic equilibrium.

Below we shall consider induced steady ordered states in some very simple open physical systems, with the aim of offering an analogous description to the one used in the last chapter for closed systems. The list of phase transitions far from equilibrium is quite long, including the instabilities of lasers, plasmas, semiconductors, and of several hydrodynamic and chemical systems. As is often the case, these phenomena were discovered independently in different fields, and it was only later that a unified view emerged by using concepts similar to those already used in the theory of critical phenomena at equilibrium.

In the following section we shall concentrate on one system, the laser, because of this system's intrinsic interest and also because it is such a well-known one that only a few introductory words are needed to recall it. Once we have understood the instability that makes the laser function, we shall have an easier time extending the notion of induced order to other physical systems.

4.2 The Laser as a Cooperative Open System

So let us consider the laser as a clear example of a physical system in a state of order induced by its openness toward the outside. Seen from outside, the laser is a tube into which goes incoherent light produced by a normal lamp and out of which comes the coherent light beam and some heat, as seen in Box 4.A. We use the term *coherence* here to mean the presence of the invariant space–time relationship which is typical of wavelike motions and is fulfilled by the electromagnetic field within the laser cavity and in the outgoing light beam. Of course, this coherence relationship is not fulfilled by the light coming out of an ordinary lamp, because the emission events that produce it are uncorrelated. Seen from within, the laser is a material system

Box 4.A

THE LASER AS AN OPEN SYSTEM

- The laser is an open system maintained in a stationary state far from thermodynamic equilibrium.

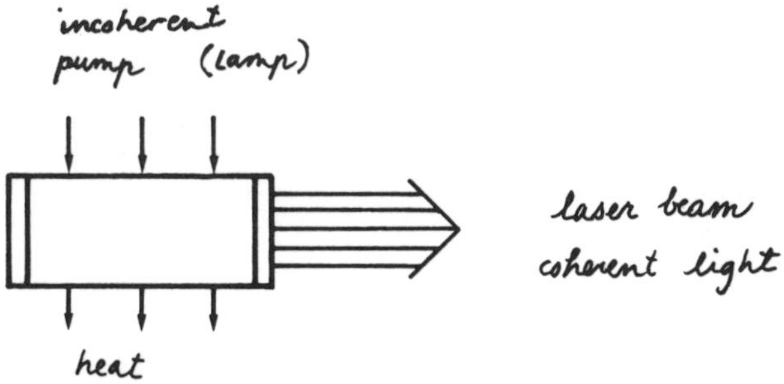

- When the power of the optical pump is higher than a certain critical value, a coherent electromagnetic wave is formed in the cavity between the two mirrors, and this generates the outgoing beam of light.

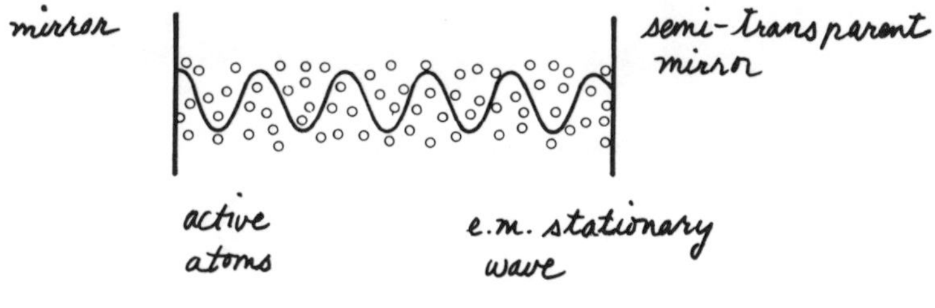

made up of several active atoms that absorb and emit light of a certain frequency and are contained between two mirrors in an optical cavity. As Box 4.A shows, one of the two mirrors is semitransparent, to allow some of the light produced in the cavity to escape. Inside the optical cavity, when the power of the light coming in from the lamp (which for this reason is also called an *optical pump*) passes a critical threshold, a steady electromagnetic wave sets in. In these conditions, the open system is kept in a stationary state of order far from equilibrium. On the other hand, the system

Box 4.B

CORRELATED EVENTS IN THE LASER

- For simplicity, we consider only one atomic level above the ground level, and use a white ball to indicate the place left empty by the electrons (black balls) during the transition, for three significant situations:

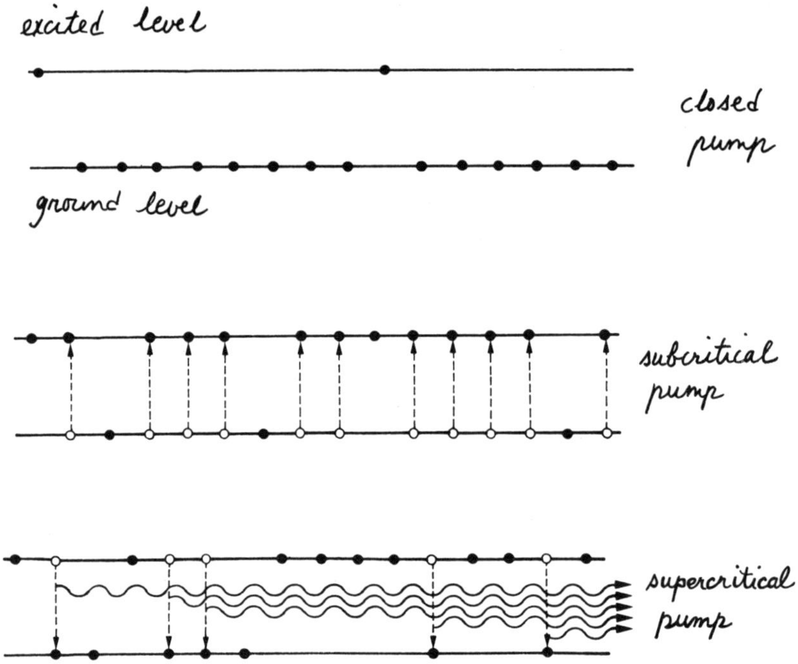

- Note that only above the critical point is the emission of a photon stimulated cooperatively by the presence of other similar events.

comes back to equilibrium when the lamp is off, but this is an uninteresting situation.

To clarify the notion of *coherence in terms of emission events,* let us consider how the laser works by looking at its energy levels and their relative populations, as shown in Box 4.B. The figure at the top of this box shows the situation when the pump is off, where the ground level is most heavily populated; for simplicity's sake, we have considered only one upper level, which is almost empty. The electrons can move from one level to another emitting or absorbing light according to the usual spectroscopy laws, but the very nature of these processes at the atomic level is of scant importance here. Of course, when the pump is on but below the critical point, the situation looks different, as has been exaggerated in the second figure in Box 4.B, where the white balls indicate the space left empty by those electrons that have absorbed energy coming from the pump. When the "critical" situation is reached, there is a cooperative effect among the transitions toward the ground state, in the sense that if an electron emits a photon, this photon stimulates the emission of another photon, and together the two particles stimulate the emission of a third, and so on. This is called the Einstein stimulated emission effect. In other words, the emission events become correlated after the critical threshold is reached.

This process of order induced in the laser can thus be described in the usual way, in terms of temporal distribution of the atomic populations among the different levels, or in terms of an appropriate correlation between the very elementary events. This last viewpoint is of great importance, because it can distinguish the transitions far from equilibrium from those that take place at thermodynamic equilibrium. Moreover, in the laser this correlation must be of a space–time nature, because of the characteristic oscillatory mode excited inside the cavity, a mode that is imposed by the dimensions of the reflecting cavity itself. But the same viewpoint can be generalized to include other types of events, such as the chemical reactions where new species appear, of which the transition from excited atom to normal atom occurring in the laser thus represents a special case. After all, a physical event is any transformation that can be observed as a change in the system's state, and it occurs when the system leaves one cell in the phase space to appear in another. Statistical mechanics has shown us that these transitions directly balance themselves in twos only under conditions of thermodynamic equilibrium, according to what is called the detailed balance principle. Now we see that far from equilibrium many events can become correlated above a critical threshold. In other words, there is a break in the isotropic symmetry governing the distribution of events at equilibrium; the formation of a new distribution is characterized by the lower symmetry proper for these events, and thus by a state of greater order.

This cooperative effect between events can also be described in the mean-field approximation, but now the internal field is certainly not spatial, because the atoms in the gas laser are not in well-defined positions. Thus we should generalize the mean-field notion to include the typical space–time correlation of the stationary electromagnetic wave. Limiting ourselves to

Box 4.C

ANALOGIES BETWEEN LASER AND PHASE TRANSITIONS

- The laser coherence is a phenomenon that may be described as a critical transition induced by an external parameter, by recording E (the intensity of the electromagnetic field of the laser beam) as a function of the power of the optical pump.

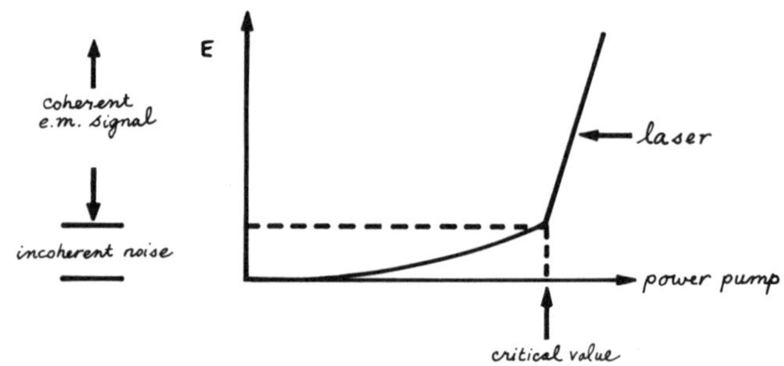

- Keeping in mind that the correlations in lasers are essentially temporal, whereas in phase transitions they are essentially spatial, we can form the analogy below:

	Order	*Disorder*
Phase transitions	The internal field creates site correlations.	Random thermal events destroy site correlations.
Laser	The internal electromagnetic field creates correlated emission events.	Random decay events destroy correlation.

the temporal aspects of this wave, which is perfectly legitimate given the high velocity of light, we may say that the electrons return to the ground state simultaneously because all these returns stimulate one another and give rise to this internal electromagnetic field. Once this is accepted, it is even possible to introduce a dimensionless cooperative parameter by combining such important temporal quantities as the decay time for stimulated emission, the normal decay time of excited states, and the escape time of electromagnetic energy from the cavity, and thus to describe the transition in terms of this parameter only. Thus it may approximately be said that just as in the critical phenomena at equilibrium the internal field creates the essentially spatial correlations against thermal disorder, now *the internal field creates the essentially temporal correlations among some events against some other random events.* The space periodicity of the binary-alloy superstructure seen in the last chapter is in a sense parallel to the temporal periodicity of the stationary wave governing the emission events in the laser cavity. In both these systems, it is not hard to see order as the relationship implicit in the constraint imposed upon the elements of the system, and more precisely as a condition of space invariance for the former or of time invariance for the latter in the distribution of matter or events.

The cooperative transition that causes the laser to work is accompanied by several occurrences that make it possible to establish an even closer analogy with the critical phenomena already seen at equilibrium, as we shall see in the next section. For the moment, it is interesting to add that the fluctuations of the coherent light intensity increase near the critical point symmetrically on both sides of it (not shown in Box 4.C). Furthermore, near the critical point the intensity of coherent light obeys a simple law of growth as a function of the displacement from the critical point so that the intensity of the electric-field vector can be taken as precisely the order parameter governing the transition from disorder to order. Finally, near the transition point, the width of the laser line decreases noticeably, and this fact clearly indicates that the system's relaxation time increases. By now, we have accumulated enough information about this open-system transition to attempt a close analogy with the critical phenomena already seen at equilibrium in closed systems.

4.3 Generalization and Analogies with Critical Phenomena

In order to be certain we understand that the behavior of the laser is also typical of other disorder–order transitions induced by external fluxes far from thermodynamic equilibrium, let us consider some other instability phenomena in the field of fluid mechanics. For example, a case of instability that has been known since the turn of the century is found in the *Bénard cell,* shown schematically in Box 4.D. In this cell, when the layer of fluid

Box 4.D

AN OPEN SYSTEM IN HYDRODYNAMICS

- In the Bénard cell, a thermal flux above the critical point generates parallel cylindrical vortices or other vortex structures dependent upon the surrounding conditions. In the example below, the velocity of a volume of fluid along the axis *z* is a sine function of the coordinate *x*, a typical feature of a coherent phenomenon.

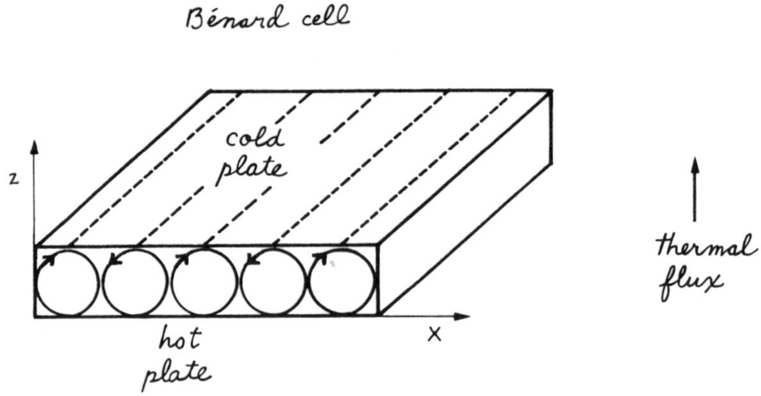

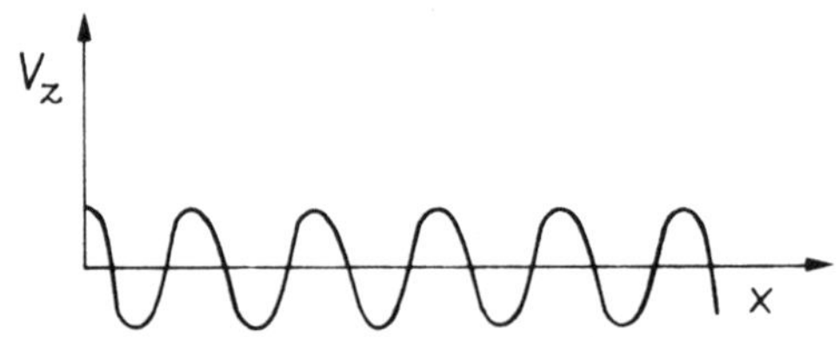

contained between one hot plate below and another cold one above is subjected to a thermal gradient larger than a critical value, a macroscopic structure appears that is heavily dependent upon the surrounding conditions. When the fluid layer is thin, this structure is made up of the cylindrical vortices seen in the drawing, and in these vortices the local velocity of the fluid is a sine function of the site with a maximum value that depends upon the gradient. Thus this is obviously a space–time ordering process, a phenomenon induced by the external thermal flux on an otherwise quite homogeneous system. The explanation of the origin of these convective structures is more complex than the one we used for the laser in Sec. 4.2, and it does not directly interest us here. Another case of hydrodynamic instability takes place among the Taylor cylinders, which are two coaxial cylinders in motion relative to each other. Here, too, for angular velocities above a critical point, periodically spaced vortical rings are created along the rotation axis in the fluid inside the cylinders. These hydrodynamic phenomena are interesting because they show that while every microscopic volume of the fluid always displays a state of local thermodynamic equilibrium, there is also the formation of an instability on a larger scale, ranging to the macrosopic dimensions of the vortex.

The examples of the laser and the Bénard cell show quite similar behavior, also typical of many other open systems. In every case we find a long-range order that originates after a discontinuity. This common behavior will be developed below using the mean-field theory, which we know is valid not too close to the transition point of the critical phenomena at equilibrium. This last fact is of little importance here, because far from equilibrium it is hard to experiment too close to the transition point. So let's see how far we can take the analogy between critical phenomena at equilibrium and critical instabilities far from equilibrium while staying within the validity of the mean-field theory.

We must first establish that *even far from equilibrium it is possible to describe the transition using an order parameter* that vanishes in the state preceding the transition. In the case of the laser this order parameter is the electromagnetic-field mode within the optical cavity, whereas in the case of the Bénard cell it is the rotation velocity along the vortical cylinder. In both cases this is a wavelike quantity that brings to mind the two-component order parameter of superfluids. Note also how close the analogy is between laser and superconductor behavior. In both cases, the order parameter is a wave function and the coherence length is macroscopic; in the first case, the external variable governing the transition is the temperature of the thermal bath, whereas in the second it is the population distribution among the electron levels, which can be formally expressed as a nonequilibrium temperature supported by the optical pump.

Fluctuations play an important role also in the open-system instabilities, just as we saw for critical phenomena in the closed systems. At the end of Sec. 4.2, we mentioned that the width of the laser line decreases as the instability is reached, while the fluctuations in light intensity increase. This means that when the instability is reached the spontaneous fluctuations that are always present in the system are amplified by the flux of the pump until they merge into one fluctuation with a long decay time. This grows at the expense of the other fluctuations, until it finally establishes itself and even takes on the permanent character of the stationary electromagnetic wave. This is a clear example of the difference that exists between the transient character of a wide-range fluctuation and the persistent character of the coherent macroscopic phenomenon that is steady after the transition.

Of course, this dominant long-range fluctuation is selected from the outside, just as in the laser the dimensions of the optical cavity determine the length of the stationary wave or in the Bénard case the cell dimensions determine the radius of the stable vortex. *Both at equilibrium and far from it, when the critical transition occurs a fluctuation is created that encompasses the whole material system and that is characterized by a long decay time.* And it is thanks to this fluctuation that the order parameter is created to govern the more ordered state that follows the critical transition.

Concerning symmetry breaking, the analogy with the critical phenomena at equilibrium is more limited. Far from equilibrium, the space- or time-reversibility symmetries are broken, simply because a definite structure is formed where there was once a complete homogeneity. Symmetry breaking in the critical phenomena at equilibrium is much richer, because of the possibilities offered by group theory, which operates at the microscopic level of the system.

However, there are some other aspects in the instabilities induced far from equilibrium that clearly distinguish them from critical phenomena. When the temperature is lowered in a system at equilibrium, a series of phase transitions take place and cause an order that must always increase according to the third law of thermodynamics. But this does not happen far from equilibrium because when fluxes continue to increase in an open system, successive instabilities are formed until a final state is reached that is usually described as turbulent. In the case of the laser, when the power of the optical pump is raised above the critical point, a new critical value causes the system to pass from a continuous wave to a series of equally spaced short pulses; next, as the power continues to rise, the system moves toward a chaotic situation of nonmonochromatic pulses, which has even been called a generalized turbulence. The same happens in the Bénard cell, where as the thermal gradient increases several instabilities are reached, each with a different type of vortex distribution, until a final instability causes turbulence. We shall return to this still obscure point later in this chapter.

Another characteristic displayed by the transitions far from equilibrium is the possibility that a system can have more than one order parameter, and that they might cooperate with each other to generate more structures. For instance, when the fluid layer in the Bénard cell is not thin, there must be three order parameters, each expressed by a plane wave function, and their combination gives rise to a structure displaying equal hexagonal cells not shown in Box 4.D. But this cooperative effect can be broken by applying even higher thermal fluxes, and then new vortical cylinders appear.

These final aspects show how hard it is to predict the structure of these open systems when we start from first principles, and it becomes even harder as equilibrium is left farther and farther behind. For this reason, we must look for a general phenomenological theory, which we shall point out in the next section.

4.4 Dissipative Structures

The theory of instabilities far from equilibrium is still in a less advanced stage than that of critical transitions at equilibrium. It must be admitted that it is truly difficult to use a unified theoretical treatment when dealing with complex systems such as those which display these instabilities in nature, and for this reason a truly microscopic approach has been possible only for the laser and for a few other oscillator systems. Consequently, a general but phenomenological approach has also been developed, starting from the hypothesis of local thermodynamic stability and then establishing a criterion for the global stability of a macroscopic system far from equilibrium. The term *dissipative structure* has been used to designate those open systems displaying an ordered state far from equilibrium precisely because they are maintained by energy dissipation. This is certainly in line with the second law of thermodynamics, because in an open system it is clearly possible to have a steady state which is also ordered, as long as an external flux of negative entropy compensates for the inevitable production of positive internal entropy, caused by irreversible effects and accompanied by energy dissipation. The hypothesis of local equilibrium requires only that entropy be defined locally on each site and not for the entire macroscopic body.

In the preceding sections we saw how order can be established in the system by forcing it away from equilibrium. The problem we are now faced with is that of finding the general conditions under which this state is stable towards spontaneous fluctuations and small external disturbances. Before going on, we should establish an important difference about the kind of externally imposed conditions: Do they induce a linear regime, in the way a temperature or concentration gradient induces irreversible phenomena governed respectively by Fourier's law for thermal conductibility or Fick's law for diffusion; or a nonlinear regime such as catalytic or feedback effects in chemical reaction chains? When the displacement from thermodynamic equilibrium is linear, a situation that is sometimes called the "Onsager regime," every state close enough to equilibrium must display stability and maximum disorder. It has been proved that time symmetry governs in this regime and that the steady state is characterized by a minimum rate of entropy production. Consequently, spontaneous fluctuations can never make the system evolve toward other more stable or more ordered states.

On the other hand, when the displacement from equilibrium is too great to be described by linear terms, the situation is different, and thus even the terms of second order must be taken into account in the entropy displacement from its equilibrium value. In these circumstances, Glansdorff and Prigogine showed that whereas at equilibrium it is the negative sign of these displacements that characterizes stability, because stability implies maximum disorder, far from equilibrium the stability criterion becomes the positive sign of the temporal derivative of this same entropic displacement. Thus it is *precisely an increasing dissipation rate that can guarantee stability far from equilibrium,* and consequently, the system must become more and more ordered.

This condition of stability far from equilibrium is quite general, but to proceed further, we must rely upon some model systems. The most suitable ones for this purpose are the chemical reaction chains that display no reaction heat because they make it easier to write nonequilibrium entropy solely as a function of the stage reached by the reaction. The forced displacement from equilibrium is performed by arbitrarily imposing the concentration value of one reaction product, for instance, the concentration of a chemical product that can be kept constant in the bath. Unlike the equilibrium situation, where the concentrations are governed by the well-known mass law, here the concentrations are quite different and depend upon the imposed conditions. In this way, we come across the interesting fact that these nonequilibrium concentrations display a temporal variation which is periodic around a steady value. In other words, a temporal structure has been created among the chemical events that originate these concentration distributions. Further computer analysis of the diffusion process taking place in the reaction box also displays some space–time structures that might well be called "chemical waves." Similar phenomena have been discovered in the laboratory for some rather complex chemical reactions, in which these space–time periodicities are maintained as long as dissipation is guaranteed by enough chemical products that are rich in free energy. However, it is still hard to say if these phenomena must necessarily result from the general stability conditions that must govern dissipative structures, or if they merely result from the nonlinear differential equations that govern the specific kinetics of these coupled chemical reactions.

Anticipating the biochemical applications to be made in the next chapter, it should be pointed out here that when an oscillatory mode in a reaction process is coupled with an isotropic transport phenomenon of the diffusive type, dissipative coherent structures can be maintained only if the dimensions of the system are large enough. This stands to reason, because when the system is too small the diffusive transport ends up destroying the coherence of the oscillatory mode. Of course, the higher the frequency of this mode, the smaller is the slow effect of diffusion. Thus we can guess that the system must have a critical size, below which exist both space–time coherence and temporal persistence of the dissipative structure. In the case of chemical reactions among simple molecules taking place in a water environment, this critical dimension is slightly less than a millimeter for oscillation periods of about one minute.

These model systems are also useful to see that certain random fluctuations can sometimes grow to reach macroscopic dimensions comparable to those of the system itself, thus selecting the instability branch that will lead to a new structure. *So it seems that chance plays a decisive role in the choice of new structures,* by taking the system farther and farther away from equilibrium in an unpredictable direction. Thus the forced evolution of the system from one new structure to another must in part have a "historical" character because of the influence of the preceding situation, but it also has a "nondeterministic" character caused by the series of bifurcations it must come across. So the evolution of an open system is also structured by the occurrence of many stationary states that are successively reached as the "pump parameter" steadily increases, keeping the system farther and farther from equilibrium. This gives the system several alternative possibilities of evolution that cannot be predicted because each branch of bifurcation is selected at random at the moment of instability. On the other hand, as the pump parameter continues to increase, the system eventually ends up in a state of chaos. How this last order–disorder transition takes place is still somewhat of an intriguing mystery, and for this reason we shall attempt, in the following section, to discuss it in its simplest context, which is that of the hydrodynamic instabilities.

4.5 From Order to Chaos

As we have seen in the preceding sections, when an external flux is imposed upon an open system, it induces a structure by means of an initial instability that breaks the space and time symmetry. On the other hand, it is also well known that high values of external flux can create a state of complete disorder, generally called the turbulent state. But there is still no clear theoretical explanation of how this inversion of behavior takes place.

It is hard to give a precise definition of turbulence, although this state of matter always displays some quite typical features. In the familiar case of a fluid, we know that turbulence is produced by rapid, nonuniform fluxes, which is to say, at high values of the Reynolds number R when the behavior of the system displays some casual characteristics not caused by external forces. Turbulence is generally a disordered occurrence, and in the case of a fluid it is expressed by the *progressive loss of correlation between the movements of its parts*. Because of this memory loss of the initial conditions, it is impossible to predict meteorological phenomena, because here the large dimensions of the fluids and the low viscosity easily lead to very high Reynolds numbers.

For a long time it was thought that this lack of predictability was caused when there were too many degrees of freedom in the system. A typical example of unpredictable behavior is the Brownian motion of a colloidal particle in an ordinary liquid, because it is caused by the resultant value of instantaneous independent collisions with the liquid molecules, a value which varies at random given the great many degrees of freedom in play. Turbulence was considered in this light, for many years, as a phenomenon caused by the great many disturbances that could develop independently in different parts of the fluid and then interfere with each other. To be more exact, because it is shown that an instability produced in a fluid in motion adds a new frequency to the hydrodynamic equations of motion of the system, Landau proposed that turbulence should be characterized as the "complicated and confused" state resulting from the overlapping of many frequencies, due to a series of successive instabilities created by ever-increasing values of the flux. As we can easily guess, this turbulent state caused by the overlapping of incommensurable frequencies should be an "almost periodic" one, and not the clearly "nonperiodic" one that is typical of a truly chaotic state. Indeed, the nonperiodic state differs from the other because of the first's complete memory loss of the initial conditions, so that two fluids that start with apparently identical initial conditions wind up evolving in completely different ways.

Today, the term *chaos* is used to mean erratic, unpredictable behavior of dynamic systems in nonperiodic regimes. But recent developments in mathematics have led to the important discovery that the dynamic equations of *systems with nonlinear interactions made up of even a few degrees of freedom* can be solved and produce chaotic solutions after a few instabilities. For instance, a coupled pendulum system becomes turbulent after a certain time, i.e., it shows complete memory loss of the initial conditions, even when it consists of only three pendula. These pendula are the three parameters included in the Reynolds number, which can thus characterize turbulence with a parameter number that is finite and not infinite, as Landau thought. The presence of turbulence in systems with only a few degrees of freedom has recently been verified in the laboratory, and this fully confirms the new conception of turbulence. For instance, in Bénard cells that are too small to contain more than a few vortices, it is possible to detect a series of successive situations, as seen in Box 4.E. After the formation of some stable vortices, there first appears a disturbance in the vortex velocity field caused by one fundamental frequency and its harmonics, followed by

Box 4.E

INTERMEDIATE STATES BETWEEN ORDER AND CHAOS

- In a small Bénard cell, it is possible to follow the stages of evolution toward turbulence, by measuring the growing time disturbance of the velocity of the few vortices present.

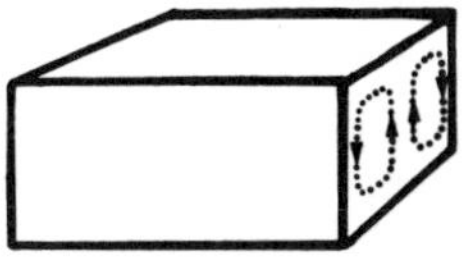

Steady regime

Periodic regime

Bi-periodic regime

Turbulent regime

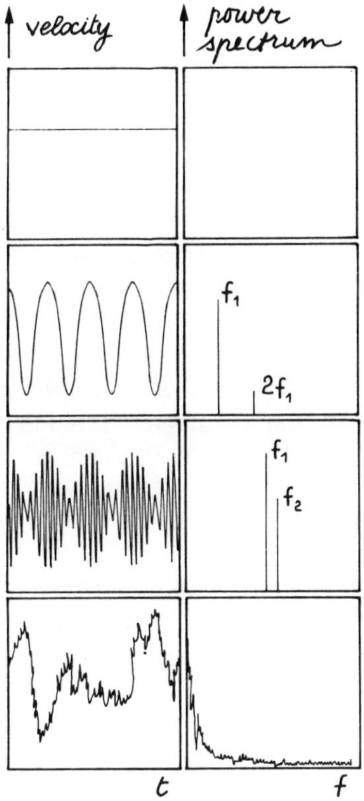

(Figure adapted from the article by Bergé and Pomeau, *La Recherche,* April 1980)

two different fundamental frequencies, and finally by the simultaneous production of infinite frequencies. Large Bénard cells behave differently; here the ordered convective state with many vortices is immediately followed by the turbulent state. Based on these most recent developments, we should say that only sufficiently "small" systems can find order first and turbulence next, whereas "large" systems become turbulent directly without first going through a more ordered state.

But we have not seen everything yet, because it appears that in many systems there is a strange regularity in the time period that characterizes the new successive state at each instability. Mathematical theories have been proposed to explain this curious property, but there is still no truly general theory that makes it possible to predict the systems that will display it. However, it cannot be denied that a great many systems behave this way, from fluids to the chemical systems, from optical cavities to the Josephson junctions, from systems made up of electrical or mechanical oscillators to the biological populations.

4.6 Summary and Conclusions

In this chapter we have dealt with the stationary state far from equilibrium, in which some physical systems open to external fluxes can be maintained. We have seen that it is possible to generate situations of stable order by raising the external flux values above a critical point. This general behavior can be described using the same concepts already encountered at equilibrium, such as the order parameter, fluctuation, symmetry breaking, coherence length, internal field, and so forth, and in addition, the notion of event.

Thus we have concluded that:

1. Space order, which we know exists in many structures in closed systems at equilibrium because of interactions inside the system, has a parallel in space–time structures supported by external fluxes above a critical threshold. The coherent electromagnetic field of the laser is the simplest of these structures, and it is the richest of these features. The cooperative phenomenon that takes place in the laser is essentially temporal, and it leads to a close correlation among emission events of stimulated light. Similar phenomena are also found in hydrodynamics, but they are harder to analyze experimentally and theoretically, perhaps because here it is so difficult to identify the elementary events.

2. Although the mean-field theory was developed to describe critical phenomena at equilibrium, it also works well enough for transitions induced

far from equilibrium. And in fact, the laser's internal electromagnetic field creates the long-range space–time correlations among the emission events that are the best expression of the order reached in competition with random decay events. The two cases differ with respect to symmetry breaking, which seems of less significance far from equilibrium. We should also add that the whole theory of order–disorder transitions far from equilibrium is still at a less advanced stage than the one at equilibrium. For instance, in the latter case some universality classes of the order parameter have not been found yet.

3. The dissipative structures supported by an external flux have their place in nature together with the structures at equilibrium that are caused by conservative forces operating inside the system. A general phenomenological theory has shown that these dissipative structures can be stable. It is also certain that long-range spontaneous fluctuations play a decisive role in the formation of one particular dissipative structure at the onset of the critical instability. This macroscopic fluctuation that stabilizes in the coherent structure depends upon surrounding conditions as well as upon chance.

4. As the external flux continues to increase, the system becomes turbulent, which means that it reaches a chaotic state formed by uncorrelated and unpredictable local states. In systems having only a few degrees of freedom, the transition toward turbulence seems to take place by passing through several well-defined successive instabilities with states characterized by a few parameters. Only when there are many degrees of freedom does the transition toward turbulence have a catastrophic character.

5. These considerations have been verified for sufficiently simple physical systems, but they appear to be so general as to be easily extended to more complex systems and even to systems that are not strictly physical. Thus for these complex situations occurring in natural processes of the greatest interdisciplinary interest, the behavior shown by the physical systems considered here is an important reference point.

4.7 References

(a) A good general introduction to phase transitions far from equilibrium can be found in the book by Haken (20). More mathematical details on the same topic are given by the same author in his article (19).

(b) The instability of the laser is dealt with from the viewpoint of cooperative systems in the introductory article by Digiorgio (10). For greater detail, see Arecchi (1). Both these articles also discuss the analogy with critical phenomena at equilibrium, more mathematical details for which are given by Graham (17).

(c) The general theory of dissipative structures is dealt with by Glansdorff and Prigogine (14); more details regarding some model systems can be found in the recent work of Nicolis and Prigogine (34). Several articles by Prigogine *et al.* (36) offer a concise introductory treatment of this topic.

(d) Today, the evolution toward turbulence is a problem undergoing rapid development, and the classical texts of hydrodynamics are completely out of date. Some recent aspects are considered in the work of Bergé and Pomeau, quoted in Box 4.E.

CHAPTER FIVE

THE FUNCTIONAL ORDER OF BIOLOGICAL MATTER

5.1 Introduction: Functional Order as a Correlation of Relevant Biochemical Events

In this chapter we shall attempt to use the notions we have developed thus far to describe the unique and highly interesting ordered state that takes place in biological processes. Since we shall consider such complex processes without going into biological detail, this chapter will be even more qualitative than the others. But our main point can certainly be made without these details; and that is, to describe functional order in the same terms used for the physical systems, namely, as a correlation among the most suitable observable quantities. Thus we shall begin by discussing the nature of these observable quantities, which are relevant to the biological processes, and then continue along the lines developed in this book by considering only the statistical thermodynamic aspect governing these processes at the molecular level. Of course, we shall limit ourselves to a phenomenological description, expressing order as the presence of some permanent relationships between the system's observable quantities. In this way, we shall avoid some fundamental questions concerning the origin and evolution of this order, because such questions can be ignored when considering the steady functional order, such as can be observed today in a cell culture in a favorable environment.

Many ways have been proposed to express the idea of order in biology. Here we shall limit ourselves to calling *biological order* the general fact that some general properties of a species remain the same, despite the differences that always exist among members of the same species and despite the continuous change of each individual throughout its lifetime. Inasmuch as the elementary processes that take place in biological matter must obey the same physical laws that govern inanimate matter, we are faced with the old problem of accounting for biological order by the laws of physics. We shall restrict our discussion of this problem greatly, concentrating only upon the nature of *steady functional order,* a term biologists use to mean that every structure is supposed to perform a function within the wide ensemble of structures and functions that have the property of self-reproduction, a property found only in the organism. However, if functional order is proposed in this light, then it, too, remains outside the laws of physics, because the teleological concept of function is alien to physical thinking. To overcome this difficulty, we shall try to describe the steady functional order as a particular correlation created among biochemical events, so as to express therein the order in which biological matter participates. In this way our attention will shift from the concepts of structure and function to that of the biochemical event, a quantity that can be observed using the methods of physics, but which is also of true biological significance.

First, let us reconsider what biologists mean by the term functional order. Biologists think of the organism as an integral unit made up of various structures and functions and capable of growing and reproducing. Thus structure and function are presented as complementary aspects of every part of the unit, with the understanding that only the ensemble of these structures has the overall function of self-reproduction. It follows that every molecular structure in the organism taken separately is essentially an organic molecule, and that it acquires its biological function thanks to the interdependence that takes place with the other molecules within the organism. Therefore biologists already consider this interdependence characterizing functional order as an ensemble of relationships that must exist among some members, which are thus ordered or, to be more precise, correlated. But now we must identify these correlated members in terms of suitable observable quantities; as we shall see at once, these quantities are the significant biochemical events.

We call a *significant biochemical event* any transformation of matter that is relevant to the biological cycle, so that all life processes of a cell can be described by a proper ensemble of these events. Every significant biochemical event is an observable quantity, because it can be measured. For instance, the synthesis of a molecule is observable, thanks to the concentration changes produced, even though the details of the transformation process are so rapid or complex that they cannot be physically observed. Obviously, a biochemical event can always be reduced to a physical fact, because the forces in play are the same ones that operate in inorganic matter. However, an event can be called significant at the biological level only if it is itself the starting point for at least one other event in the series of events that gives rise to cell duplication.

It must be understood that a series of events can have a true biological significance only if it is composed of the exact number and type of members that allow cell duplication. Thus it is the entire series of events that makes a single event significant, but it is the presence of every single significant event that allows the existence of the successive events of the series in this closed loop. This concept will be clearer if we think of the series of events that takes place in the very simple case of a metabolic cycle, where each event depends entirely upon the previous event. In this case, the probability that an event may occur at a given time can be expressed as a function of the time lag only between this event and the previous one, by means of a time correlation in one variable, which is the time lag.

Since every event is connected to others in one or more permanent relationships, such as the simple temporal correlation seen in the last example, *we may identify the functional order as the complete ensemble of correlations existing among significant biochemical events.* These correlations may actually be quite complex mathematical functions, they may be spatial as well as temporal, and they may contain more than one variable if more than one preceding event generates the event in question. But the crucial point is that now this ensemble is made up of quantities that can be observed physically because they are biochemical events, so that the order in which biological matter participates can be formally described in the same way as that of inanimate matter, namely as a correlation among the most suitable quantities.

However, the formal description of the order thus reached in these two states of matter does not imply a true analogy with the nature of correlation displayed by the different cases. For instance, in the case of a crystal lattice, the spatial order is best expressed by the presence of correlations among the positions of equal atoms, whereas in functional order the correlations must be formed among the times at which different events occur. In the case of the crystal lattice, this order is further characterized by a condition of invariance toward the space transformations allowed by the symmetry class of the lattice in question. But in the case of functional order, it is hard to formulate mathematically the condition of invariance that must be fulfilled, which can be stated broadly as the need to keep the characters of one species constant during all transformations that give rise to biochemical events.

Proceeding in our comparison between physical order and functional order, we see that whereas the coherence of a laser is characterized by the presence of a wavelike space–time correlation among exactly similar events, namely, the optical emission events stimulated within the optical cavity, biological functional order must be characterized by time conditions which cannot be so rigid. This is because although the general plan of significant events is materially written in DNA as genetic information, the actual performance of this plan as real events depends upon some random environmental conditions, since the environment controls growth by supplying matter and energy. Consequently, inside this DNA-planned invariance each event actually takes place according to temporal features that may even be of a random nature, and this must be more and more true when events occur at the molecular level. Thus we come to the important conclusion that *the time correlations among significant biochemical events that occur at the molecular level must be of a statistical nature.*

The spontaneous biochemical event seems to be the most suitable element to describe order at the level of cell life, inasmuch as here life is unseparable from the growth process that takes place in space and time, and because these are the proper coordinates that characterize a physical event. However, we must realize that the description of functional order is much less intuitive than that of physical order because it is so hard to formulate concretely the correlation among so many different events. This limitation is only partly due to the impossibility of keeping track of all the significant events that occur in a cell and to the fact that these events sometimes involve structures hard to observe using the methods of physics. The real obstacle in describing functional order as an ensemble of correlations among significant events lies in *the intrinsic complexity of these correlations.* Since just this irreducible complexity is an essential character of these correlations, we should consider this point more closely here.

This intrinsic complexity arises because the same structures are often used in different functions, so that some events become connected in parallel as well as in series. Let us take an example to explain this important point. Inasmuch as ATP is the currency used for all the metabolic reactions that require energy support, it follows that there is a reciprocal control among the reactions occurring in the same region, because their rate depends upon the concentration of one compound shared by all of them. The same is true for some complex structures that display different functions, as we shall see in the case of biological membranes. So one is faced with a large group of events linked to one another by a multidimensional network which, when observed from the outside, seems of an insurmountable complexity. But only this multiple correlation can guarantee *the reciprocal control for necessarily statistical events* and so make it possible for one cell to operate as a unit. Thus the complexity seen by the observer is yet another expression of the correlated ensemble of events that characterizes functional order, but it is not easily describable, as in physics, by the introduction of an order parameter. This is an intrinsic complexity that prevents the reduction of the biological system to simpler physical systems.

This state of complexity has led to two different scientific approaches. Biology has attempted to offer a qualitative description of some structures and functions in terms of other, more elementary, ones. But others have been trying to formulate a possibly quantitative reduction of the most elementary biological events to real physical facts. This is the method we shall use here to describe local functional order, following the same conceptual lines as those developed in the preceding chapters. In the next section we shall try to make these introductory concepts clearer by using the concepts of correlation among biochemical events and of polyfunctionality of a complex structure for a case in which these concepts can be more easily visualized, namely the local functional order found in biological membranes. Unfortunately, I am afraid this explanation will be useful only to those who are already familiar with the essential phenomenology of these membranes.

5.2 Local Functional Order in Subunits: Biomembranes as Event Correlators

As we have said, the production of a significant event at the molecular scale of the organism depends upon a suitable statistical correlation among previous events. However, this correlation cannot be fully expressed, because of the complex network which includes that event. But if we limit ourselves to a sufficiently small part of the cell, it may at times become possible to express this correlation in detail, so as to be able to describe the local functional order in that part of the cell. As a matter of fact, by partitioning space into appropriate compartments, nature eliminates some interactions and thus those correlations that might be dangerous or unnecessary. For this reason, we shall now consider the biological membrane, a small structural unit capable of performing several functions, in order to see if we can now describe local functional order in terms of correlation among significant events. In short, we shall find that the membrane is an organ that can provide *the correlation among some primary events, to lead to subsequent events by means of a selective process.*

The biological membrane is an extremely widespread structure found in all organisms, and it displays at least three kinds of major functions. In the first the membrane separates two compartments, such as in the cytoplasmic membrane surrounding the cell or the nuclear membrane enclosing the nucleus of the cell, and thus it works as a semipermeable barrier that regulates the flow between the two compartments. In the second, it supplies a surface where complex events can occur at a higher rate, such as in the mitochondrial membrane and the endoplasmic lattice. In the third, it provides a threshold process for some critical molecular or ionic concentrations, such as in the neuromuscular junctions of vertebrates. Of course, there are many other, more particular, functions, but there is no need to mention them here; and there are some other complex functions we shall mention later. While in nature there is no single chemical structure able to perform all the above-mentioned functions, one finds several varieties with a similar chemical structural design, to fit the primary function to be performed. Thus the morphology of the membrane lies between two models that correspond to the extreme cases of a static and uniform structure or a dynamic and non-uniform structure, both formed of an appropriate lipid and globular protein mixture. Without going into further detail about the chemical structure, we can now see how the three main functions mentioned above can be expressed in terms of correlations among some specific events.

As we have already mentioned, one of the first functions of the membrane is to *provide vectorial and selective fluxes* through the surface of a compartment. Obviously, thermodynamic equilibrium would require that the concentration on each side of the membrane be the same for every molecular species, and nature does not oppose this "passive" transport caused by molecular diffusion, which tends to eliminate every gradient. However, nature can add only an "active" transport that costs energy but is capable of bringing something back to its original position. By this active transport, a high concentration of molecules is kept in the compartment and this allows for some chemical reactions that would otherwise have a negligible yield. Now let us try to reformulate this well-known picture in terms of correlation among appropriate events. To the observer watching the inside of the membrane, active transport looks like the appearance of molecules on that surface and thus like the production of events that are space-correlated. In an artificial membrane under conditions of equilibrium, only random fluctuations can occur in the number of events, which on the average is close to zero, because the principle of detailed balance states that every event must be compensated by the opposite event. But this is no longer true in a biological membrane, thanks to the active transport, and so it is possible to achieve this space correlation among events, given that their average number is no longer zero.

Now let us consider the second function of the membrane, when it works *to facilitate kinetic processes along its surface.* Nature tries to reduce space dimensions in order to increase the rate of chemical processes that rely upon diffusion, when reagents are available only in low concentration, in the same way two people are more likely to meet if they are moving along a road rather than in a field. In this way, an enzyme attached to the membrane can greatly increase its collision cross section toward a rare molecule moving on the membrane surface instead of in the free space; furthermore, if the overall reaction needs several enzymes to work in series, their proper strategic distribution on the membrane can give a much greater kinetic yield. This may all be reformulated by saying that the high probability that a fairly rare event will take place is attained thanks to a specific space–time correlation imposed upon some primary events. This is made possible by the microscopic structure of the membrane, a structure which has been selected by evolution for this very end.

Now let us consider the membrane's third function, which is to *give a threshold response to the concentration of certain molecules or ions* (called *effectors*). At first it was thought that this property was caused by the cooperative transition of several units called protomers, interacting in a field within the membrane, like spins or atoms in a quasi-rigid two-dimensional lattice. But later developments gave greater importance to a membrane model that is almost fluid, in which protomers are mobile, like icebergs in the ocean. In this case, the cooperative action must be found inside the protomer itself, which is a complex globular protein made up of different subunits set in the membrane structure. It is thought that these subunits can act in a concerted way, after the effector acts upon one of them, and that something of the sort can actually take place is shown by the *in vitro* behavior of hemoglobin, which we shall see later. So in this case, too, the action of the membrane may be described as a facilitated correlation induced upon biochemical events, if we identify the elementary event as the effector-operated transformation of one protomer subunit, and the correlation as the cooperative effect among the subunits embedded in the membrane.

Thus we may conclude that each of the three main functions attributed to biological membranes can be seen as the formation of a certain type of facilitated correlation among events. At this point, note that these three primary functions take place in the same spot, thereby making it possible to give rise to other *secondary functions*. For instance, as Box 5.A shows, the active transport generated by some different proteins on the same side of a membrane can cause the appearance of different molecules, which can next react with each other thanks to the subsequent action of the enzymes also attached to the same side of the membrane. It is clear that these more compound functions cause secondary events that may be expressed as correlations among primary events. Again, the enhanced probability that an otherwise very rare event can occur because of other events can be expressed as a correlation function among these preceding events, and again, this correlation may be spatial or temporal. But if the membrane displays too many interconnected correlation functions, the observer may end up seeing this membrane as a single complex unit, because it is operatively impossible to separate, say, the actions of the protein that provide transport from those that help catalysis, and thus it may be impossible to formulate those correlations that express the local functional order.

The example of the membrane is also useful because it shows how unimportant a mere structural notion of order can be when describing a biological function. Indeed, the lipid membranes first attracted the attention of some physicists because they were thought to offer the possibility of linking the change in space order that takes place in their phase transitions with some truly biological properties. Although it is certainly true that some biological membranes grow in conditions close to their phase transitions, so as

Box 5.A

CORRELATION OF EVENTS IN BIOLOGICAL MEMBRANES

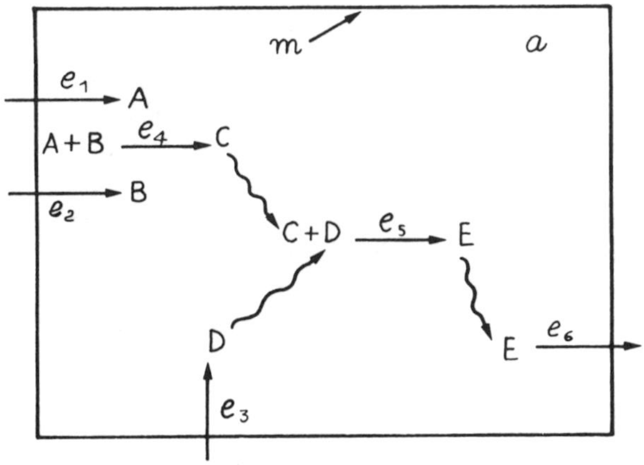

A schematic example of events *e*, correlated in a compartment *a* enclosed within a biological membrane *m*. The input of the molecules A, B, D, E is supported by active transport through the membrane and can be represented as the simple events e_1, e_2, e_3, e_6. Because of the local spatial correlation among these primary events, secondary events are next formed: the e_4 type directly on the membrane helped by the enzymes attached there, or the e_5 type in the internal liquid phase. The final product E then leaves the compartment, whereas the intermediary product C cannot get out.

to reach a fluidity that makes fixing and moving easier for embedded proteins, it has also been found that structural-type phase transitions have no general biological relevance. This is because what really matters for the living body is the functional, rather than the spatial, order of the structure, and the correlation among significant events rather than among space sites. And in fact, a biological end is always achieved by means of selection, and this selection is obtained at the molecular level by a particular facilitated correlation among primary events, which thus generates the secondary event necessary to functional order.

Biological membranes are cell organelles that work with events at a molecular scale, and consequently the facilitated correlation they induce must have a statistical nature. If we move up through the hierarchy of functional order, we come to other organs like the muscle, in which this correlation among events becomes more and more rigid until it turns into a true causal relationship. But we shall leave these macroscopic aspects to return to the molecular ones, which are more in line with the character of this book.

5.3 Programmed Events in a Thermal Bath: Enzyme Action

In this section it will be our concern to see how the most elementary but significant biochemical events involved in the functional order can originate. Inasmuch as these events occur at room temperature at the molecular level, kinetic disorder cannot be avoided, and we may ask ourselves if this disorder does not interfere with the programmed character that the growth process must have. To be more precise, how can the microscopic reversibility of random events be reconciled with the irreversible and selective character of the programmed events? And how can such events occur spontaneously? The answer to these questions can be found in a closer examination of the selective action mechanism of the enzymes which, thanks to their particular structure, are capable of producing programmed events by making proper use of disorder itself.

In the first place, we should realize that the action of the enzymes must take place with the greatest speed so that new structures of the organism will stand the competition with the inevitable decay processes. It is true that all the operations in the chemical processes of a cell do occur with an amazing rapidity. For instance, a bacterial cell can completely reproduce itself in about twenty minutes, and during this time a genetic message containing about four million letters must be read and put into operation through enzyme synthesis. Thus one can calculate that the mean time for an elementary biochemical event must be close to a millisecond. Therefore, the programming of these significant events must include the very short time they take to occur, as well as the specificity of the events themselves.

Nature has solved the problem of this rapid and faithful generation of significant primary events in a thermal bath by using macromolecular structures that have been adapted to this purpose by evolution. If this generation is possible, it is thanks to the flexibility of some globular proteins, composed of chains with several thousand strategically arranged atoms that permit the best sequence of internal events. For our purposes here, we shall call internal events those relevant events that take place inside the macromolecules and into which the chemical reaction process can conceivably be split, while we shall call the global event the final and permanent change observable in the reaction products. Then we can say that thanks to the sequence of internal events, the reaction pathway is optimized to yield the global event. As we have said, this last event—the synthesis of the chemical product—is truly significant if it becomes the starting point for other events of that series which leads to cell duplication. Without the appropriate enzyme the generation of this chemical product would be such a rare event that it could never become time-correlated to other events, and so the whole ensemble of correlations that lead to growth would vanish. Actually, an enzyme is essentially a catalyst, and if we want to understand its action we must first consider the chemical transformation of matter that occurs in every catalytic process.

Let us look at the simplest chemical system, one composed of two species A and B, each of which transforms into the other according to the scheme

$$A \rightleftarrows B$$

inside a closed isothermal box that also contains the catalyst. By definition, the catalyst can speed up the process only toward equilibrium in both directions. Now let us suppose that the concentration of A is greater than the one required by thermodynamic equilibrium, so that the observer sees the inevitable production of B in the box. *Thus the catalyst appears to be a generator of events,* because almost no event would be observed in the same time scale without it. So thanks to some reversible microscopic operations it is possible to generate events only in one direction if the environmental conditions allow it. This consideration replies to our above question about the apparent irreversibility of the organism's chemical machine.

Now let us try to reach an idea of the physical process that must take place in catalysis. If the A + B system still has not reached equilibrium in isothermal conditions, it must have been inhibited by a free-energy barrier. Thus the action of a catalyst consists of lowering this barrier to the point where some thermodynamic fluctuations of free energy can overcome it so that the transition from A to B can take place. The lowering of the highest point in the barrier, which is called the transition state, can be best obtained by making a sophisticated network of chemical bonds, which include all reaction partners and even the catalyst. In practice, the catalyst must link the reacting molecules in some favorable positions and next, it must deform these molecules until they flow through the transition state which has the lowest free energy. The best catalyst is the one that can force the reacting system through many configurations to reach the one that can be overcome thanks to a thermal fluctuation. In this way, the system can go beyond the transition state. The conflicting needs of protecting the reaction site from outside disturbance and linking it to the thermal bath, which is the source of energy, make the design of a catalyst quite a difficult task.

Having established this much, let us see how nature has solved the problem of the best catalytic action by using globular proteins. Biochemists propose to describe the enzyme action as a succession of steps composing the reaction pathway that transforms the substrate, the name given to whatever must be transformed (see Box 5.B). The interactions involved are electrosta-

Box 5.B

SEQUENCE OF EVENTS IN ENZYME ACTION

- In the simplest case of an enzyme E that must catalyze the dissociation reaction of a chemical bond of a substrate S, the enzyme action is split into three separate steps:

1. The molecule attaches itself to the active site of the macromolecule, forcing it to take on a complementary position.

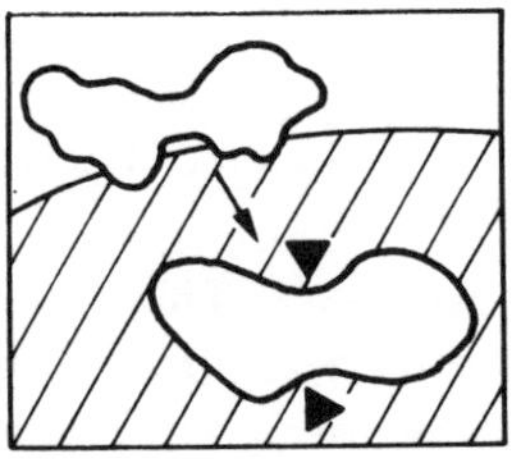

2. Substrate and enzyme form a single unit, in a fairly rigid transition state X*, from which the system goes on to evolve by forming the two products of the chemical reaction.

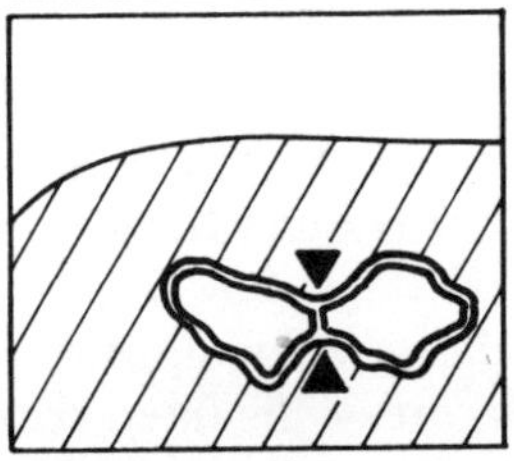

3. The reaction products leave the active site and spread throughout the surrounding bath.

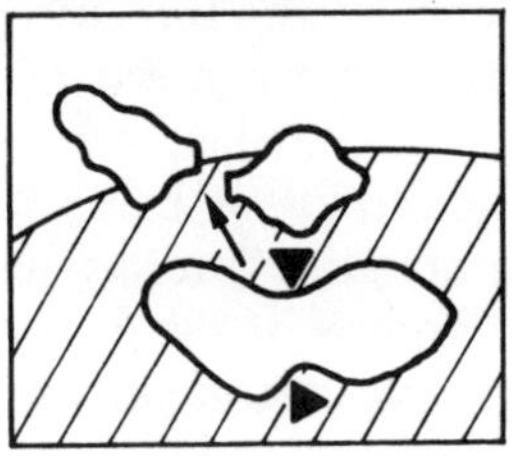

- From the thermodynamic viewpoint the sequence of the three reactions above can be considered as three consecutive activated processes (starred*) below:

$$E + S \rightleftharpoons ES^* \rightleftharpoons ES \rightleftharpoons EX^* \rightleftharpoons EP \rightleftharpoons EP^* \rightleftharpoons EP + E$$

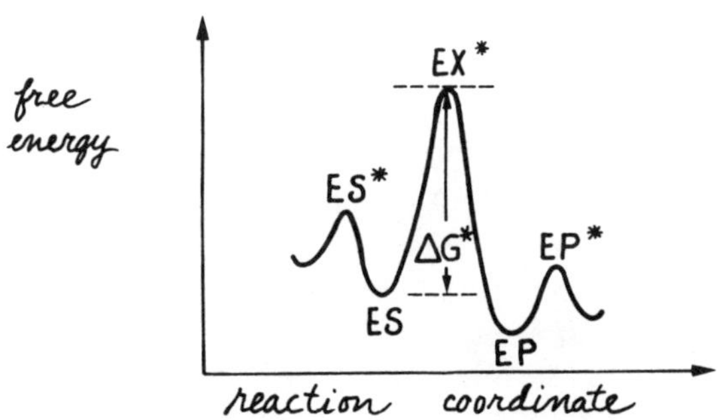

tic, and often quite weak in comparison with the thermal energy, so that only *the concerted effect of a great many interactions* can take the system, consisting of the substrate plus the enzyme, to the transition state of each step of the reaction pathway. This point is important, because it shows that both thermal disorder and structural order are necessary to form a relevant internal event such as the crossing of a transition point. As we have said, the global event is the result of a series of these internal events in macromolecules. The first internal event is to link the substrate to a sterically convenient site; next some changes in the configuration of the macromolecule must occur to optimize the active site where the transition state must be built and then overcome. This optimization implies the correct orientation of the active groups and the formation of steric barriers to create the best transition state, the one of lowest free energy. Consequently, the free-energy lowering produced by the enzyme is essentially entropic; therefore a transition state should be thought of as a kind of ordered structure, even if not observable. As we can see, the enzyme must behave both as a structure that is generally quite flexible but also as a structure that can become temporarily quite rigid. The flexible nature of this structure under the thermal agitation allows the temporal sequence of internal events, whereas the cooperative nature of the interactions gives rise to the particular order of atoms that characterizes the transition state.

To go further in our microscopic description of enzyme action, without going into biochemical details, it would be convenient to envision the macromolecule as being made up of quasi-rigid subunits called domains. These domains interact weakly among themselves and are bonded to the active groups of the common polypeptide chain of the protein. So we see that the most strategic distribution of the active groups can be reached by random fluctuations of the reciprocal positions among domains and substrate. Conversely, what happens is that the fluctuations of all the polypeptide-chain sections within the domains are drastically reduced, because these disturbing fluctuations would make it take much longer to reach the transition state. This is all shown in the sketch in Box 5.C. Note also that if the macromolecule were completely rigid it would be impossible to have that sequence of internal events that are necessary for the chemical pathway. Thus evolution has given the macromolecule a design that utilizes both thermal disorder and spatial order, *a design toward a programmed goal that is reached in spite of some randomness.* In other words, spatial structure has been optimized so that chance may play its role in the course of time. Thus thermal disorder is an energy-carrying factor, whereas the macromolecular structure is a limiting selective factor, and both are essential to the programmed production of events.

At this stage it would be interesting to note how nature has made systematic use of a particular bond, called the *hydrogen bond,* to form all the

Box 5.C

THE ROLE OF ORDER AND DISORDER IN ENZYME ACTION

- The increase in free energy $\Delta G^* = \Delta U^* + T\,\Delta S^*$ in the transition state X* drawn in Box 5.B is largely due to the negative entropic contribution $-\Delta S^*$; this makes it possible for the active groups, which are bound to the rigid domain structures of the enzyme, to find the correct positions.

- Thus both the spatial order of some enzyme parts and the disorder of others are essential to its action, and both types of part belong to the same polypeptide chain, in a design that optimizes their function.

- This schematic drawing shows how the two domains D_1 and D_2 can arrange themselves to be productive versus the substrate S, thanks to the random action of some hydration-water molecules (black balls) and of the disordered part of the protein polypeptide chain, which induce a Brownian movement on the whole rigid domain.

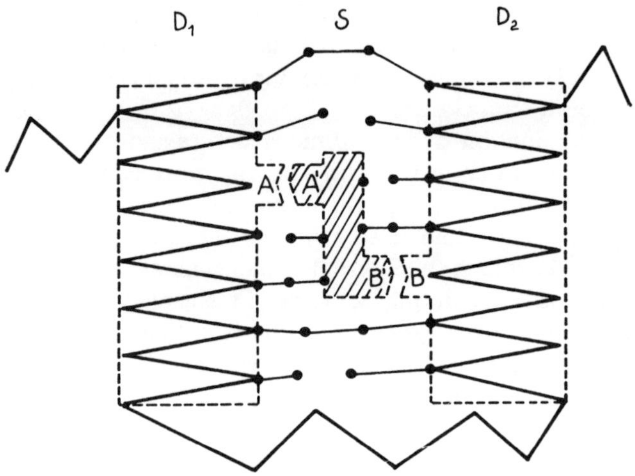

functional units that must be partly rigid and partly mobile. This is usually a weak bond, in the sense that the two bonded partners are usually linked by an energy that is not far from the thermal, but this bond can even become much stronger when the two partners are brought closer together. For instance, the double helix of DNA and the polypeptide chains of globular proteins are maintained by many weak hydrogen bonds that can be formed and broken cooperatively. On the other hand, at the active site of an enzyme, these bonds may be shortened, and thus they occasionally become quite strong. Even water, which makes up about 80 percent of cellular matter, is a polymer made of continuously changing hydrogen bonds. And for this reason, because of the many hydrogen bonds existing with the surrounding water molecules, the globular proteins are deeply affected by the thermal disorder of the solvent, and they must continuously change their shape.

Thus, the hydrogen bond is the ideal element for the formation of structures that must change at room temperature. This also shows up in the experimental fact that the free energy of formation of the globular proteins is close to the thermal energy. This is due to an almost exact compensation between entropic and enthalpic terms of the free energy, a fact that is typical for structures made of many hydrogen bonds, and must imply the presence of high-amplitude thermodynamic fluctuations in these structures. *It is thanks to these spontaneous fluctuations that events are generated inside the enzyme,* the sequence of which leads to the global event. A many-body system linked by hydrogen bonds, such as a macromolecule with its water solvent, is thus the most suitable matrix for continuous creation of molecular events at room temperature.

The x-ray examination of an enzyme structure shows how versatile the polypeptide chain can be in forming both the more compact sections that are inside the domains as well as the more mobile ones that are in contact with the solvent. This chain also contains the active chemical residues, which play a direct role in the internal events generated by the fluctuations in the relative positions among domains. This example clearly shows *the role played by the genetic information* that is present in the sequential arrangement of the active residues of the polypeptide chain, as well as *the role played by chance,* which acts through the flexibility of the globular protein. Nature uses these contrasting elements of order and disorder to put together the series of generally unobservable internal events that lead to the global observable event, which is the permanent transformation of the substrate into the product of enzyme reaction, and which in virtue of its programmed nature falls into the realm of the cell's functional order. We cannot help but note how the series of time-correlated biochemical events that characterizes the cell-growth process can be found, as it were, within the elementary act of the chemical transformation that is performed by an enzyme, inasmuch as even this act seems to be made up of a sequence of internal events statistically correlated in space and time. This last space–time correlation has not yet been observed experimentally in a compact globular protein, because its domains are too closely interdependent, but it seems to emerge from what we have just seen. For this reason, it would seem more sensible to look for this correlation in a globular protein made up of separate chains, which we shall do in the next section.

5.4 Polyfunctional Macromolecules at Equilibrium

At this point, we shall try to see if a time-correlation can possibly be formed among the internal events of a macromolecule at thermodynamic equilibrium. The best case to study is hemoglobin, which is of interest because its behavior is typical of a whole class of polyfunctional macromolecules that are widely used by the organism to *keep events at the molecular scale under control.* The internal event in these macromolecules can be identified in the operation of binding a small molecule called a ligand (in the case of hemoglobin, an oxygen or carbon monoxide molecule). When a ligand is already bound to a spot on the macromolecule, this makes it easier to bind other ligands and thus we may say that one internal event controls those that follow it. One interesting group of such polyfunctional macromolecules is displayed by the allosteric enzymes, whose name stresses the indirect effect performed by the ligands on the global operation of the enzyme upon its substrate, thus by activating or inhibiting this operation.

Hemoglobin is the protein responsible for transporting the molecular oxygen required by metabolism, and for removal of the carbon dioxide produced by the respiratory process of the higher animals. One molecule of hemoglobin contains about ten thousand atoms divided into four separate polypeptide chains, each having a sequence and shape resembling myoglobin. Thus each chain is itself a globular protein, a compact subunit, and four subunits joined together in a tetrahedral arrangement to form the functional unit of the hemoglobin molecule. Each subunit contains a heme group, made by a porphyrin ring in which oxygen can be bound to an iron atom. The four spots at which the four oxygen molecules are located are separated from one another by some protein sections, so that there can be no direct interaction among these spots. On the other hand, the affinity of each spot toward oxygen depends upon the binding at the other spots, so it can be said that ligand binding is a cooperative effect. It should be added that other processes, which do not affect the heme group, can influence the affinity toward oxygen; for instance, when hydrogen ions are bound far from the heme group they cause oxygen to be released. All these effects have important physiological consequences, because in this way hemoglobin can transport oxygen from lung capillaries to muscle tissues, where the partial pressure of oxygen is much lower than in the atmosphere, and where it can be released thanks to local changes of the pH level.

Box 5.D

COOPERATIVE EFFECTS IN HEMOGLOBIN

- Hemoglobin is made up of four subunits that resemble a myoglobin molecule. The affinity of hemoglobin to oxygen increases gradually as oxygen is bound to the subunits, thanks to a cooperative transition that takes the whole unit from a relaxed state R to a tense state T.

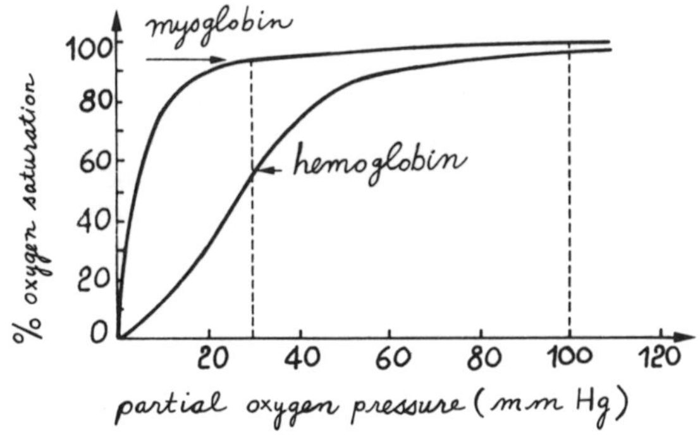

- The binding free energy of hemoglobin versus oxygen is different in the R and T states, which are in thermodynamic equilibrium when two oxygen molecules are bound, so that the oxygenation process marked by the arrow is easier as the amount of bound oxygen rises.

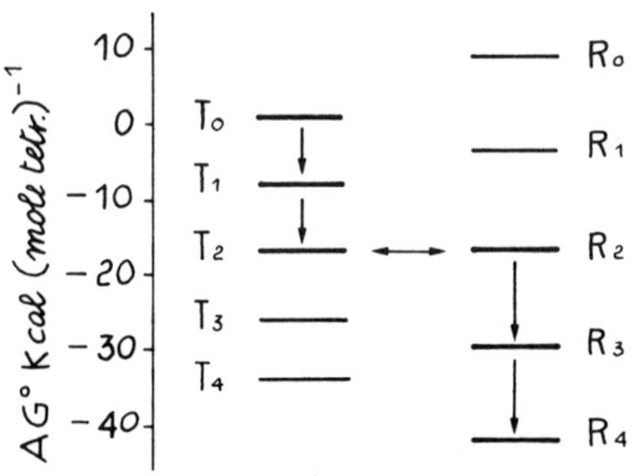

In the first place, let us consider the thermodynamic aspect of oxygen–hemoglobin binding. Today it is certain that the entire hemoglobin molecule can exist in *two slightly different conformational states,* one a relaxed state R and the other a tense state T, which have different affinities toward oxygen. This situation is seen in Box 5.D, which shows that the conversion from one state to another can take place freely when two levels coincide, and this happens when two oxygen molecules are bound to some of the four subunits. Consequently, following the arrow drawn in the box, when the first two oxygen molecules have been attached, a T→R transition takes place which increases the binding energy toward the other two oxygen molecules. The opposite effect occurs in the inverse process of dissociation, in which the first two molecules have a harder time separating from the subunits whereas the other two have less difficulty, thanks now to a R→T transition. It should also be pointed out that in this process there is no direct interaction among ligands, because the state energy levels are equally spaced. These thermodynamic aspects lead us to the sure conclusion that this reversible cooperative R⇄T transition among the four subunits makes it possible to *statistically correlate the internal events in groups of two* within the global event, which consists of binding four oxygen molecules to the macromolecular unit.

Now one might ask if the kinetics of oxygen binding also follows the same trend as that of the thermodynamic affinities and thus can likewise be expressed in terms of a cooperative transition among the four subunits. If it can, then there would also be a two-member temporal correlation, because the mean time required to bind oxygen at the two empty spots would depend upon the moment when the other two spots were first bound. We may infer that this space–time correlation among internal events should be caused by the properties of the four polypeptide chains that participate in the R⇄T cooperative transition. This is our question, which may be generalized by asking *if the internal events of a macromolecule composed of subunits can be time-correlated by means of concerted changes in the conformation of the subunits.* Unfortunately there is still no sure answer to this important question, because as yet it is impossible to detect this time correlation among subunits in the laboratory. There is an indirect way to prove the existence of this kinetic cooperativity in the case of hemoglobin, by means of an *ad hoc* analysis of the kinetics of the system, but there are no direct and conclusive data yet. Thus while it is certain that there is a static cooperation, the kinetic cooperation is only likely but unfortunately still hypothetical; furthermore, it may not be required by the biological function in the case of hemoglobin.

The microscopic details of the static cooperative process have still not been completely identified, and this is quite understandable given that we are dealing with slight local changes in highly complex structures. On the other hand it is certain that the $R \leftrightarrow T$ transition can occur thanks to the globular protein structure's flexibility at equilibrium. To be more exact, the $R \leftrightarrow T$ transition of hemoglobin takes place with quite small changes inside the four polypeptide chains, and with greater changes in the reciprocal positions among chains. Thus it would seem possible to extrapolate the same reasoning to enzymes made up of a single polypeptide chain composed of several domains, namely to the case we considered in the last section. Then it would follow that thanks to the conformational changes induced by a substrate in one domain, it should be possible to inhibit or facilitate the entire catalytic process even in the reversible realm of the Onsager regime.

Even the static cooperative nature of the conformational transitions of the macromolecule at equilibrium can thus give rise to *a mechanism of positive or negative amplification of the events,* in the sense that the first event in the series can facilitate the successive events, or it can inhibit them entirely. In the case of hemoglobin, we have seen the particular instance of two equal events that make two other equal events more likely, but it is clear that very many possibilities can be available to polyfunctional macromolecules. Incidentally, note that the water molecule is also a ligand, and that the external surface of the macromolecule can bind several hundred hydration-water molecules, so that the molecular properties of the solvent may play an important role.

Because ligands can come and go to and from different parts of the cell (both far from and close to the polyfunctional macromolecule in question) the latter can control and likewise be controlled by some remote events in the cell thanks to ligands that freely diffuse. Considering the wide diversity of the events that must be kept under control, we might expect that the nature of these controls would differ greatly from case to case. But nature has found the most general control element in globular proteins, by making good use of their *conformational flexibility at thermodynamic equilibrium.* In other words, by using diffusion processes and thus still inside the Onsager regime, globular proteins can control and thus correlate molecular events even for a rather wide range of distances, but of course with rather long time constants. For shorter times, control can only be achieved by making use of processes that are far from equilibrium, as we shall now see.

5.5 Biochemical Dissipative Cycles Far from Equilibrium

In the preceding section, we saw how the cooperative conformational transitions of allosterical molecules allow the control and thus the maintenance of a steady state of order among events by working near thermodynamic equilibrium. Now we shall try to see how the intrinsic nonlinearity of the cooperative response can even lead to instabilities in the biochemical cycle where these macromolecules are located, by creating even a long-range order, an authentic coherence that is typical of the dissipative phenomena we met in Chapter Four. As we already know, this coherence must be supported by a steady energy supply in an open system, and this condition is certainly satisfied in the life cycle of a cell. Now, it happens that the utilization of ATP in the recharging process of high-energy chemical compounds is always mediated by some allosteric enzymes, and thus this control generates dissipative effects. So the concentration of chemical compounds can display a wavelike distribution, which implies correlations among the events that play a role in the cycle where this coherence is maintained. Let us try to clarify this notion at once by looking to a concrete example, *the glycolytic cycle.*

Glycolysis is the process of dissolving sugar, and this term is used to mean the most widespread mechanism that allows microorganisms and even animal cells to use glucose anaerobically for energy and nutrition purposes. Glycolysis consists of the transformation of glucose into lactic acid, with simultaneous production of ATP from ADP and of phosphate P. Respiration is a similar but more complex process, where the energy yield is even higher, because the final product is not lactic acid but carbon dioxide and water. At any rate, the glycolytic process is complex enough in itself to require a chain of at least 11 enzymes that operate in series and are dissolved in the cytoplasm, unlike the enzymes involved in respiration that are suitably fixed in the mitochondrial membrane. The enzymes of the glycolytic cycle are complex and separate units, but the whole process is designed in such a way that the optimal yield is obtained by controlling only one particular enzyme of the chain, in this case phosphofructokinase (PFK). The way this enzyme PFK operates is greatly influenced by both ATP and ADP, in the sense that ATP is produced much faster if ADP is already present at high concentration, whereas this production is slower if instead ATP is present at high concentration. Inasmuch as the sum of the ATP plus ADP concentrations must always be constant in the cell, their ratio is thus kept near the optimal value. In other words, ADP is an activator, while ATP is an inhibitor for the reaction that produces ATP itself.

Now let us take a closer look at the PFK macromolecule, which is an *allosteric enzyme* made up of eight subunits, each with a molecular weight near 90,000. This enzyme acts like hemoglobin to bind several positive and negative effectors besides ATP and ADP. Although most of the molecular details of the cooperative transition are still unknown, we should expect this enzyme to respond to the effectors in a nonlinear way, and thus the network in which this element is located should display an oscillatory behavior. So the concentration of the intermediate chemical products of the chain should display the oscillatory behavior characteristic of the open chemical systems in a dissipative regime. Indeed, these oscillations have been experimentally detected in the millisecond region, in both intact yeast cells and cellular extracts. Therefore, we can have no doubt that some of the periodic activities of biological systems may even be caused by one allosteric enzyme set within a sequential process of events. As we saw in the last chapter, these time phenomena consist of dissipative structures that change the distribution of concentration, with all those wavelike features typical of the long-range coherence that can so easily be simulated by computer. In particular, this simulation shows that the joint presence of biochemical oscillations and spatial diffusion makes it possible to transmit chemical signals at macroscopic distances with high velocity and low attenuation. This possibility still has to be verified within a single cell, but similar phenomena have been found in colonies of microorganisms, where periodic AMP pulses produce both the simultaneous development of different cells and their successive aggregation. Thus we should also expect that some fundamental biological events such as mitosis and cell differentiation could be caused by some oscillations generated by an allosteric control of the elementary events. In this way, the time correlation of some events can extend well beyond the range of molecular interactions, which instead determine the value of the kinetic constants of the chemical reactions and the diffusion coefficients.

The possibility of building up *a long-range correlation with a short time constant,* characteristic of the coherence displayed by dissipative effects, has another important implication worthy of attention. As we already know, far from thermodynamic equilibrium the relationship between chemical kinetics and space–time concentration distribution can no longer be described by local terms only, because the solution of the kinetic equations also depends upon such macroscopic factors as the size and boundaries of the system. So we must expect this last dependence to be decisive in overruling the possibility of reducing long-range space–time order to short-range effects alone. Inasmuch as this consideration concerning matter distribution must raise a similar implication for the event distribution, it follows that no rigid determinism can control these events, either.

These comments thus show the importance of the external energy supply, not only to keep an open state far from equilibrium, but also to produce new situations which may then be optimized and stabilized by evolution. For this reason, the dissipative structures are sometimes thought to be responsible even for the genesis and evolution of functional order itself. But because of the irreducible complexity we have already mentioned, it is still impossible to identify a suitable order parameter, representative of the functional order of an entire cell, well enough to describe the growth process as the formation of dissipative structures in simple physical systems. It is still hard to assess how essential dissipative structures are for the present state of functional order, because we know that a steady state in an open system can be self-controlled even in the Onsager regime, in the realm of microscopic reversibility. In other words, once the statistically time-correlated series of programmed events that leads to cell growth has come to exist, this steady process near equilibrium can repeat itself in favorable environmental conditions. Furthermore, while the only advantage of the dissipative structures is to be able to allow some controls with short time constants over long space distances, the biological necessity for this type of control in a live cell must still be experimentally proved.

5.6 Summary and Conclusions

Besides displaying the forms of order and disorder considered in the other chapters, matter also takes part in functional order in life. We have tried to examine this form of order that is peculiar to biological matter in the same way we did for inanimate matter, expressing order as a correlation among suitable observable quantities. Along these lines, we saw that:

1. The event, and not the "structure–function" binomial, is a better quantity to describe life phenomena, inasmuch as they cannot be disassociated from the growth process, which occurs in space and time. The significant biochemical event is the elementary act by which biological matter is transformed. This even may be unobservable throughout its course, but it can always be observed in the new chemical product it generates.

2. Functional order, the term biologists use to mean that every structure has a function within a unit of structures and functions capable of growing and reproducing itself, can thus be described as the ensemble of statistical correlations among some significant biochemical events. This description is completely analogous to the one for physical order, since it is always expressed through appropriate space–time correlation functions. But this is a methodological viewpoint and not a realistic program of work, because it is impossible to observe all the events that occur in a cell and because there are so many connections among the various processes. This means that the complexity of the system to be observed is irreducible. We cannot say that these correlations are the manifestation of an order parameter that governs this open system far from equilibrium.

3. If we limit our discussion to the local order shown by a given complex structure, it may be possible to identify some details of the time correlations among events. This was done for the example of the biological membranes, which operate as correlators of some primary events to selectively produce other secondary events.

4. The most elementary biochemical events of functional order are generated by the enzymes, macromolecules that have been selected by evolution

to make otherwise extremely rare events very likely. The elementary biochemical event is itself the result of a series of internal events that occur within the macromolecule, and that generally cannot be observable experimentally. The enzyme's flexible structure is designed in such a way as to use both the structural order of some of its rigid parts as well as the thermal disorder induced by the solvent. The spatial correlation needed by the internal event is made possible by a large number of weak interactions that are present in the macromolecule and seem to act cooperatively. The sequence of internal events is made possible by the flexible nature of this structure, which fluctuates around its equilibrium configuration.

5. Hydrogen bonds play an essential role in biological matter, because of their property of generating both the structural order of the rigid parts and the random fluctuations among their relative positions. These wide fluctuations around the mean equilibrium configuration are caused by a compensation between entropy and enthalpy that brings free energy very close to thermal energy.

6. The static correlation among internal events can be more easily traced in the macromolecules made up of separable subunits not strongly coupled. This is the case of hemoglobin and many other polyfunctional globular proteins used in the indirect (allosterical) control of some elementary events, which usually consist of binding or releasing small molecules. This correlation of events is achieved at equilibrium by means of cooperative transitions among the subunits, which require some concerted conformational changes. But the presence of a kinetic kind of correlation, even in the reversible Onsager regime, is more uncertain.

7. What is certain is that through the kinetics of molecular cooperative effects, still unknown because of the complexity of the many-electron system, thermal disorder and structural order in macromolecules must combine to produce the elementary biochemical events later involved in the stationary functional order on a wider scale.

8. The nonlinear response of allosterical enzymes, caused by the cooperative nature of their transitions, can also generate instabilities in the biochemical cycle to which they belong, if the process is maintained by a supply of external energy. This is precisely what happens in the most important metabolic cycles such as glycolysis, the example considered in this chapter. Because of the dissipative effects generated in these open systems, a coherent macroscopic distribution of matter may form and this calls for a long-range correlation of events with shorter time consultants than those of the merely diffusive phenomena. These correlations must be unpredictable,

because they are caused by the stabilization of a random fluctuation over dimensions comparable to those of the system. This unpredictability can be of great importance in the evolution of the biological system.

9. Thus it seems that at least at the molecular level the notion of correlation among events can be used to describe functional order for usual physical systems. The genesis and evolution of this order call for other considerations that fall outside the confines of this book.

5.7 References

(a) A general introduction to biological order is given by Lwoff (28), whereas Lehninger (27) deals with some more biochemical aspects.
(b) The notion of the statistical biochemical event is treated by Careri *et al.* (8).
(c) Biological membranes are described in every textbook of molecular biology. Some recent physical aspects can be found in an article by Nagle *et al.* (33).
(d) An introduction to the flexibility of enzymes is given in the article by Koshland (24); the statistical aspects are treated by Careri *et al.* (8). For other biochemical details, see Koshland and Neet (25).
(e) Further general information on hemoglobin can be found in the article by Perutz (35).
(f) The dissipative aspects of biological structures are treated in some detail by Hess *et al.* (21), and more generally by Prigogine *et al.* (36).

APPENDIX A

LOCAL ORDERED STATES IN THE EVOLVING UNIVERSE

At every level, the structure of cosmic matter reveals the existence of evolutional processes. There is no doubt of this, and the visible signs range from the Earth's crust to the life cycle of a star. All of these processes share one essential quality: Matter is transformed from a simple and uniform state into another, more complex and differentiated one. Everywhere, cosmic matter displays what has been called the arrow of time or, to be more precise, *the historical arrow of time,* to distinguish it from the opposite process shown in all irreversible phenomena of which we have experience, the process known as the *thermodynamic arrow of time.* Whereas the historical arrow of time moves in the direction of increasing information at the macroscopic level, the thermodynamic arrow of time goes in the opposite direction, with a loss of such information. How can we reconcile the apparent contradiction between these two time arrows?

Let us briefly recall the statistical reason for the thermodynamic arrow of time. Information has meaning only if it concerns properties of the system that imply some statistical values. Thus there is also information at the microscopic level, and it is related to our knowledge of the system's properties at the atomic scale, such as a correlation among the atomic velocities. In the irreversible processes that obey thermodynamic laws, information is transformed from the macroscopic level to the microscopic one, and it is there that it is lost. This is one way to express the second law of thermodynamics, which we know to be valid only for a closed system. Actually, as Borel showed in 1912, the information loss at the microscopic level occurs precisely because no physical system can be considered truly closed, in the sense that it is the interaction of the particles with the container walls that destroys the correlation among particle velocities and thus leads to a loss of information at the microscopic level. In other words, the thermodynamic arrow of time exists because *no real closed system is a truly isolated one,* and thus macroscopic information is transformed into microscopic information with the consequent loss of the latter type, because of the inevitable disturbances caused by the world outside the system.

However, the situation must be different for the universe, considered as one body, because only in this case are we dealing with an isolated system. Here if there is microscopic information, it cannot be lost, because all interactions occur within the system. However, there can be no such microscopic information, because of a fundamental property of the universe, namely its isotropy, to which we shall return below. Therefore, we see why it is absolutely impossible to extend the usual thermodynamic argument on the arrow of time to the universe.

The problems raised by considering the universe as one body are quite complex and cannot be treated without general relativity. But our concern here is with astrophysics and not with cosmology, and so we would be wiser to formulate our question as follows: How can ordered structures be formed in such large regions of the universe that they extend almost to the limit of our present means of observation? To emphasize the fact that we wish to consider only a part and not all of the universe, we shall call this order a "local" phenomenon.

Our discussion of this local order can only proceed if we first recall a few things about the evolution of the universe itself. In the first place, since Hubble's observations in 1920, we have known that galaxies move reciprocally away from a common starting point following a simple law. This led Gamow to formulate the "big bang" theory of evolution, which says that when it was a few minutes old the universe must have been an exceptionally dense and hot mass, a "fireball" composed of radiation and completely ionized gases. In this phase, the radiation energy, which can be expressed as a mass using Einstein's equivalence between mass and energy, greatly outweighed the material mass present. This phase was followed by an expansion, which lowered the temperature, and then opposite ions recombined and the electromagnetic radiation was able to escape into free space, leaving matter behind in a situation like the one we see now. We should point out at once that the expansion of the universe must mean that there is a cosmological arrow of time, and while this fact is of great philosophical import, we shall accept it here as a mere experimental finding. But another result of general importance is that the observed expansion leads us to accept a model for the universe which in Einstein's cosmogonic theory requires what has been called *the cosmogonic principle,* which says that the distribution of matter and its state of motion are statistically homogeneous and isotropic quantities, with the exception of some local irregularities. The final proof of this universal isotropy came in 1965, when Penzias and Wilson observed a radiation background in the microwave region, which reaches the Earth with an angular constant of a few parts per thousand. This radiation background has a frequency distribution that corresponds to a black body at 2.7 K, and it was precisely this radiation that Gamow had predicted as the present-day residue of radiation in his evolutional theory of the big bang.

After these few words, let us now try to reformulate our question in even more definite terms, as follows: How can ordered structures take place during the expansion of the universe, being true irregularities in light of the cosmogonic principle which instead requires universal isotropy? As we shall see, theoretically it is quite plausible to have such local irregularities, but it is much harder to justify the particular structures that have been observed, unless we add some assumptions to theory.

Given that we are interested only in local aspects of the universe, let us try to work from a thermodynamic viewpoint to make some predictions about *an expanding closed system made up of radiation and heavy, nonrelativistic matter.* We shall recall that in the well-known case of a gas expanding in a cylinder limited by a moving piston, the thermodynamic transformation that causes the gas to cool off can take place without producing entropy, but only if it is infinitely slow. This is the only way to avoid the irreversible processes connected with the inevitable delay between piston motion and gas molecule motion, disregarding the friction of the piston against the cylinder walls, which does not interest us here. But when nonrelativistic matter and electromagnetic radiation expand together, some temperature and pressure differences must certainly take place, because the state equations of the two subsystems are different. The inevitable coupling of the two subsystems reestablishes internal equilibrium, but if this relaxation time is longer than the expansion time, the relaxation process must take place far from equilibrium, and the entropy of the entire closed system must grow. Now, it is obvious that the relaxation time depends upon the temperature and density of the two subsystems, but particularly upon the reciprocal collision cross section among its members, because it is always a collision mechanism that establishes equilibrium. Consequently, it is hard to say more about the entropy increase in a closed system in expansion without specifying which system we are discussing. For instance, the interaction between radiation and matter differs greatly depending upon the state of ionization of the matter, because of the extremely high collision cross section between radiation and free electrons in the ionized state.

Actually, the situation is complicated even further by the effect of gravity, which locally increases the density of matter, thus producing an increase in temperature and pressure. The greater information resulting from the increased density of matter in a given place when a structure is formed should then be considered alongside the increased entropy caused by heating the mass itself. But this heating up may be lowered by radiation if the system is open toward the rest of the universe, so the optical properties of matter control outgoing radiation and thus allow the formation of a permanent structure if the system can cool off.

This reasoning should make it clear that *during the expansion of the universe states of order and disorder may be formed locally,* according to the extent of the region under consideration, its connection with the rest of the universe, and the local conditions of matter and radiation. And because matter and radiation have been in a continually changing situation since the big bang, we should expect to find a variety of ordering and disordering processes in the different regions of the cosmos. Given this introduction, we shall now try to treat some topics in greater detail, within the framework of all that we have seen so far, by examining two extreme cases of local order.

A.1 Cyclical Processes in Stellar Evolution: The Cepheids

A cyclical process is essentially a particular kind of correlation among events, and thus it always expresses a space–time order.

Let us sidestep the formation of the first stars and consider some cyclical processes that involve all the stars, as we can observe them today. The thermonuclear chain reaction that takes place inside a shining star is a cyclical reaction, but we shall not consider it here because it calls for a knowledge of subatomic matter. However, there are at least two other types of cyclical processes that do interest us. The first is the great cycle of evolution that involves all galactic matter in a statistical correlation consisting of a continuous exchange of matter between stars and interstellar clouds. Moreover, in the lifetime of a star there are also some true periodic cyclical phenomena maintained by irreversible processes; the one we shall consider below is the remarkable case of the Cepheid pulsations, which can act like true galactic clocks precisely because of their dissipative structure.

In the galaxies, stars are continuously being generated from clouds of interstellar gas composed essentially of hydrogen; they often end their lives violently in the same galaxies, when the thermonuclear reaction has burned off all available fuel. Before the thermonuclear reaction is set off, the cloud of interstellar matter is not in a state of thermodynamic equilibrium. For reasons resembling those recalled in the last section with respect to gravitational instability, it has been shown that even this small cloud, subject to internal Coulomb and gravitational forces, has no true state of equilibrium characterized by a state of minimum internal energy. Consequently, because of inevitable local fluctuations, the system evolves with time, in the sense that some of its particles are expelled and others are concentrated at a lower energy level. The initially homogeneous interstellar cloud is thus split into an external region that becomes colder and colder as it expands more and more, and an internal region that is hotter and hotter as it contracts more and more. It is in this internal region that the thermonuclear reaction is set off when a certain temperature is reached, changing the chemical nature of matter, and later causing this matter to be emitted into the surrounding interstellar space because of a further increase in temperature. This emission of matter generally takes place through an almost continuous expulsion of sprays of incandescent matter, and sometimes through the explosion of the entire star, as happens with supernovas.

It is the matter emitted from the stars during their active life that gives rise to new stars, seemingly because the shock wave generated by the spray disturbs the density of the interstellar gas always present in the galaxy. We realize that there is a *steady cyclical process involving both the stars and the interstellar cloudy matter* in a slow intertransformation, at the same time that a final noncombustible residue, namely the central kernel of the spent stars, forms that we shall discuss in the next section. Of course, since the amount of the nuclear transformation residues present in the cycle steadily increases, we should also expect a slow change in the kinetics of the elementary processes that maintain the cycle.

Now let us consider the lifetime of a star in evolution. Many features (e.g., its duration and the amount of residue) will differ, according to its mass, but we can best orient ourselves in this topic by concentrating on the characteristic time constants of this evolution. The shortest one is the relaxation time a pressure wave takes to propagate through the star. In the initial phase, this time governs the condensation of the gaseous mass, whereas in the subsequent phase it controls hydrostatic equilibrium (a few hours, in the case of the sun). Next, there is a characteristic thermal relaxation time, also called the Kelvin–Helmholtz time, which governs the heating of the gaseous mass during the compression caused by gravity (10^6 years, in the case of the sun). And finally, there is the time involved in the thermonuclear reaction, which greatly depends upon the mass in question (10^9 years, in the case of the sun). Because these three times differ greatly, these irreversible processes should not be able to be coupled, and we should expect the star to have a monotonous life that might be described as the slow and continuous decay of all its properties, including its luminosity. This is not always true, as in the case of the Cepheids, where we actually see a cyclic pulsation of luminosity lasting several hours. The existence of such a regular phenomenon superimposed upon the irreversible evolution of the star must imply a kind of coupling between some irreversible processes, as encountered in the previous chapters. And, in fact, we shall now see that this is the case.

A normal star does not pulse, because even if the equilibrium between the pressure of the gas heated by the thermonuclear reaction and the pressure caused by the force of gravity is perturbed, the system reverts to equilibrium by dissipating its excess compressional energy as heat. This process is characterized by pressure waves that decay exponentially with a proper time constant. The situation is analogous to the well-known case of the compression and rarefaction of a gas in a cylinder, where the imperfect thermal insulation of the walls leads to damping the otherwise perfectly periodic temperature and pressure changes of the gas. Consequently, when a star is at equilibrium, the temperature and pressure of the gas cannot change periodically without being damped, and this is directly displayed by its luminosity, which depends upon temperature and thus cannot show regular pulsations.

The phenomenon of perfectly regular pulsations observed in the Cepheids can be produced only when there is also another process that allows the system to be "recharged" at the end of each phase, in order to overcome the inevitable damping, just as the spring of a clock keeps the system moving in the presence of friction. In the Cepheids, the recharging mechanism is set off by the gaseous helium present in the outer layers of the star. Helium is ionized more easily by radiation when compressed, and it restores this energy in the form of heat when it expands, thus avoiding a progressive cooling off. To be more precise, in the first phase ionized helium absorbs the radiation energy released by the thermonuclear reaction taking place at the center of the star, and in the next phase it gives it back as the recombination heat of a reaction between ion and electron. This is a good example of how the coupling of two irreversible processes, one caused by damping thermal oscillation and the other by a photochemical dissociation followed by an exothermal recombination reaction, can lead to a process that seems to be coherent only because of the kind of observation we are making here. We do not observe the coupled irreversible processes that sustain the phenomenon, just as in a clock we do not observe the friction process, but see only the periodic motion of the hands. The particular structure of the active stars containing a great deal of helium thus makes it possible to observe the *perfect isochrony sustained by an energy flux coming from the internal thermonuclear reaction.* The analogy with the periodic phenomena observed far from equilibrium and mentioned in the preceding chapters is thus complete. Of course, as always in astrophysics, it is impossible to influence the phenomenon, and thus it is impossible to follow these instabilities through all of their phases, as for some physical systems discussed in Chapter Four.

A.2 The Regions of Greatest Order in the Universe: White Dwarfs and Neutron Stars

From an experimental viewpoint, the final state of matter at the center of a dying star is quite similar to the one we can reach in the laboratory at the lowest temperatures. In dying stars, this state of extreme order is caused by the exceptionally high pressure, which can compete with the opposite effect caused by the extremely high temperature. Current theories state that, when the nuclear fuel runs out, a star may become a black hole, or a white dwarf, or a neutron star, depending upon its dimensions. We shall not consider the first possibility, which occurs only at more than two solar masses, because we could not say very much at all without using the theory of relativity. Instead, let us try to observe the two other possibilities, which can be discussed along the lines we have already seen in the foregoing chapters.

Theory predicts that stars weighing less than about 1.4 solar masses at the nuclear-fuel exhaustion stage must contract because of gravity, until they reach dimensions close to the Earth's, but with an internal density of at least 10^8 g/cm^3. This is possible because the high pressure brings the atoms so close to one another that the electrons leave their atomic orbits and thus create a sea of electrons, which in such pressure and temperature conditions is extremely degenerate. As we know, in a degenerate Fermi gas, the less volume is available, the higher must be its zero-point kinetic energy. It is just this kinetic energy of the electrons that can compete with the pressure generated by gravity, causing a situation of stability. Thus this situation can be reached only in sufficiently small dying stars, which in astronomy are known as *white dwarfs.*

The idea of an ordered structure that should be found in dwarf stars was first proposed by Fowler as long ago as 1926, only one year after the Pauli principle had been formulated. Fowler's daring picture stated that matter in these stars should be considered as one gigantic molecule in its lowest quantum state, whose great internal energy cannot be spent in radiation, as would happen in an atom or a normal molecule. Indeed the zero-point energy of the degenerate-electron sea cannot be removed from the system, just as it is impossible to "stop" an electron in the atomic ground level. But the atomic nuclei present in the star must also display a crystalline type of order, because a crystalline structure is generated by the equilibrium of the compressional forces (which in this case are always gravitational) and of the repulsion forces (which are obviously of a Coulomb nature and are due to ions of the same sign, which pressure has caused to lose their electrons). Of course, the lattice cell of this crystal must be much smaller than the one usually found in metals. But at this point we shall omit further detail regarding the structural geography of the white dwarfs, which depends upon chemical composition, temperature (close to one hundred million degrees), and distance from the center. It is clear that such a strong dependence of the state of matter upon density means that the structure must display quite a marked inhomogeneity going from the center toward the crust of the star; here we have considered only the center of the star, where matter is homogeneous and pressure is the highest.

For stars whose mass is between one-tenth and about twice that of the sun, there is another possibility, which is to end their lives as *neutron stars.* This occurs when matter is compressed until its density reaches 10^{14} g/cm^3, the density inside the atomic nucleus. Consequently, equilibrium must now be maintained between the pressure generated by gravity and the zero-point kinetic energy of the nucleon sea, made up essentially of neutrons. Because of their extreme density, these stars should have a radius of only about ten kilometers. They are thought to have been identified in the pulsars. They are likely to take the state of highest order that one can predict for matter over such a large region, for reasons we shall now see.

First of all, let us see why the neutron sea originates. When the density is even greater than the quite high one of the white dwarfs, the electron velocity increases until it approaches the speed of light, because the smaller the available volume the higher the zero-point kinetic energy must be. But the kinetic energy of the electrons is now so high that the nuclear collision between electrons and protons produces neutrons. When the density reaches 10^{14} g/cm^3, there are almost no more free electrons, and only neutrons and a few protons can remain. Now thanks to its zero-point kinetic energy, it is this degenerate neutron gas that is able to overcome the pressure generated by gravity, just as the electrons in the white dwarfs did earlier, because both electrons and neutrons obey Fermi statistics. But in the neutron stars, there is also the possibility that the sea of neutrons will behave like a superfluid, as we saw for mass 3 helium, because both mass 3 helium and the neutron are Fermi particles with similar masses. Thus *inside a neutron star it is possible to achieve the most ordered state of matter* ever produced in the laboratory at the lowest temperatures, even though the temperature of the star is close to one hundred million degrees. Even the protons, which account for only a small percentage of the number of neutrons, can be in a superfluid state, and because they are charged they must be able to keep the electrical currents caused by their motion persistent, as do superconductors.

The proposed identification of rotating neutron stars as the pulsars should make it possible to observe in the universe some of the typical phenomena of superfluids seen in the laboratory at the lowest temperatures. If the matter inside a star is in a superfluid state, the rotation of the star is not due to the rigid rotation of its whole mass, but can actually be achieved only thanks to some quantized vortices. Just as for a rotating cylinder of liquid helium we saw the presence of a lattice of parallel quantized vortical lines, so for a neutron star it is predicted that there should be a similar structure with vortical lines 10^{-2} cm apart and with an internal radius of 10^{-12} cm. The oscillation mode displayed by this vortical-line lattice must be of peculiar value for this stellar matter, with some characteristic frequencies thought to be present in the radio signals generated by the pulsars.

The few electrons that still remain in a neutron star must also be in fairly abnormal condition. They may well be trapped within the quantized vortical lines. Although they cannot give rise to a superfluid, not being dense enough, their velocity is so close to the speed of light that their kinetic energy becomes greater than the energy of their rest mass. At even greater densities, these electrons must give rise to a series of nuclear reactions that produce elementary particles, as in the laboratory with electrons of this energy obtained by the most powerful accelerators. But now, because of the very high density, these elementary particles should not decay, and they should give rise to a state of matter that is still unpredictable both theoretically and experimentally. This unknown state of order, the greatest of any we can conceive of today, may already be present in the central kernel of the neutron stars.

In conclusion, it may be said that a low-temperature, laboratory-inspired model for a neutron star should be made up of a spherical iron shell filled with superfluid helium and having a small central region of unknown matter. Given the very small amount of normal fluid present in the neutron sphere, the corresponding liquid helium in the laboratory should be at about 10^{-2} K. At the temperature of 10^8 K, the inside of a neutron star seems to be the place for phenomena analogous to those that take place at the lowest temperatures in the laboratory, and it seems amenable to the same conceptual framework already used for superfluids.

APPENDIX B

THE IDEA OF ORDER IN SCIENCE

The idea of order is as old as thinking itself. To understand means to produce some sort of order among sensory impressions by creating appropriate general concepts, forming relationships among these concepts by means of a syntax, and then comparing the predictions made on the basis of these relationships with later sensory experiences. Natural sciences have refined common thinking to a purposefully limited part of the world around us, the one that can be measured, and it should be no surprise that the idea of order plays a leading role here. This idea has taken on more and more specific aspects as science has had to organize itself into more and more disciplines as it has grown; nonetheless, order still retains its original meaning, which we might identify as *the persistence of common relationships in a wide class of entities.* To proceed with this identification, we should concentrate on three scientific disciplines: (a) mathematics, where the only limitation to thinking is the need to give a logical basis to the quantitative relationships between freely chosen entities; (b) physics, where these entities are perceptions of the outer world, as expressed in metric units (and we can certainly reduce to physics all the other disciplines that work with more complex forms of inanimate matter, such as chemistry, geology, and so on); (c) biology, which for now should be kept separate from physics because of the peculiar character of the order displayed by living matter. Our first aim here will be to analyze the idea of order separately for each of these three scientific disciplines, so that we may then return to scientific thought in more general terms.

B.1 Mathematical Order

Since the beginning of civilization, mathematics has concentrated on the notion of measurement, which is to say the possibility of representing the properties of an object, such as its weight, in abstract form using a number. Thus the science of numbers has revealed itself to be the grounds for objective knowledge, by making it possible to identify some elements and distinguish them from one another. But when the notion of number is introduced, this also implies a sequence among the elements. The spontaneous appearance of this sequential relationship among the numerical values that represent the measurements of a group of objects immediately offers the idea of ordering these objects in a series of progressive numbers that represent these measurements. This is the idea of *sequential order* that initially emerges from mathematics. However, whereas ancient thinking concentrated essentially upon the notion of measure, progress has brought mathematical thinking to refine the very notion of order, which is now recognized as the basic notion common to all aspects of mathematics.

The modern idea of mathematical order originated from that of the sequence of data, but it was then generalized to the idea of a common relationship existing in the ensemble of all the elements considered. Today, in mathematics a particular element becomes meaningful only after analysis of the way the ensemble's elements are systematically related to one another. Likewise, the mathematicians who formulated the *ordinal number* theory a century ago did not think that order was merely a property of the number, but considered order the real principle able to characterize that number. These mathematicians did not imagine the number to be a self-sufficient entity that could later become a member of an ordered ensemble. For them, the single number was no more than *a particular site in a general ordered structure,* and it was this structure which determined the meaning of the number. Thus they transformed the meaning of the number from an absolute to a purely relative quantity. In 1893, Dedekind said that the fundamental philosophical notion upon which the whole ordinal-number theory is based is only "the mind's ability to refer certain things to certain others," and using this function alone he showed how it was possible first to define the sequence of natural numbers, and then to create the zero, then the negative, fractional, irrational, and complex numbers.

Thus mathematics is and always will be the science of an order considered as a system of relationships. When it deals with forms, it is never concerned with their essence, being interested only in the relationships that may be stated between these forms. The single elements are determined and thus receive their meaning only from the systematic connection that orders them. Each element is defined by means of the others, and none is independent of the others. In 1892, Papperitz said that "the object of pure mathematics consists of the conceptual relationships that we can establish among some elements imagined in any way we like, if we consider them a part of a suitably ordered ensemble, and that the law according to which this ensemble is ordered must depend upon our choice.'' One kind of ''geometry'' can be nothing but one particular system of relationships whose character is established by some operational laws. Thus in the development of modern geometry, purely projective thinking contradicts the metrical one, and in 1822, Poncelet conceived a geometry based exclusively upon the notion of relationships between positions and no longer upon the ideas of form and measure.

The idea of order in mathematics has also been developed by introducing the notion of *invariance,* which has played a remarkably important role in all the sciences. The simplest way to introduce this notion is to imagine a figure designed on a sheet of rubber, which can then be folded or pulled any which way without being torn. Now let us ask which property of this figure has remained unchanged during this transformation. It is clear that the values of the angles, the lengths of the lines, or the area of the figure have not remained the same. The only thing that has, and thus the only thing that can be called invariant, is the order with which the different points making up every line follow each other. This remarkable *persistence of one property during a certain type of transformation* is developed thoroughly in group theory, to establish the notion of *symmetry*. We broadly use the term symmetry to indicate the existence of several viewpoints from which the system looks the same. The connection of symmetry to invariance and to order can be understood by thinking of the persistence of the identity of a square, despite all the rotations of a right angle around its center. What remains the same is the order followed by the points that make up the sides of the square, and these points are the elements of the ordered ensemble being considered.

Intended as the presence of permanent relationships between the elements in an ensemble, the idea of order can again be generalized to include the *statistical order* that may exist among some random variables. For instance, when a die is tossed, six different numbers may be produced at random, and each has the same probability, which is one-sixth. There is an important theorem of probability calculus that proves, for a large enough ensemble of these random events, the existence of a well-defined distribution law for the probability of such events. The existence of such a law again implies order, because it implies a relationship among the entities considered, which in this case are the probabilities of a random event.

In conclusion, the idea of order is truly fundamental in mathematics, and it always suggests an interdependence among the elements of any ensemble we can imagine. It is interesting to point out here that mathematics does not consider disorder as a self-standing notion, but only as the lack of order.

B.2 Physical Order

Physics tries to offer a self-contained quantitative description of natural phenomena by using simpler and simpler and ever more general laws. Thus physics does not propose to explain nature, not even inanimate nature. It is interested only in the *regularities* displayed by the simplest events that can be measured in the laboratory. These regularities express a type of order that acquires greater consistency, thanks to a process of progressive synthesis, which leads first to the formulation of a particular law to express the regularity in a restricted group of events, and then arrives at the formulation of more general laws to express a common feature of a wide class of particular laws. We shall use the term *physical order* to refer to this well-defined system of relationships fulfilled by natural phenomena, so as to distinguish it from mathematical order, which would instead encompass all the relationships we could imagine to exist between the measurements of these natural events. Here we shall point out only incidentally that there is no logical process that can lead from the raw data to the formulation of the more general laws that are often reached through intuition.

A physical law always expresses a correlation among a group of observed events. For example, the law of falling bodies can be considered to be a correlation among the space–time events included in the sequence of positions taken by the body, and the time when each position is measured. Once the initial conditions have been given, nature chooses only one of many possible correlations, because of the value of the gravity constant, and of the laws of dynamics. These laws were formulated by Newton in his search for a common character displayed by all these types of terrestrial events and also by another group of events observed in the motion of the planets around the sun. Thus classical mechanics makes it possible to express the general order displayed by nature in the self-consistent relationships existing between natural events. In quantum mechanics the fundamental laws may be directly formulated as correlations among successive observations made on the same object, observations which are now the events themselves. Thus physical order can always be expressed in terms of correlations among events and as further relationships displayed by these initial correlations.

Following Wigner, we may reformulate and enlarge our previous considerations by saying that just as the observed events are the raw material that becomes ordered by a particular law of physics, these laws are themselves the raw material that can become ordered and expressed by the so-called *general invariance principles,* in the process of progressive synthesis mentioned above. We know that a general invariance principle is equivalent to a *symmetry principle,* and that it generates a *general conservation law.* For instance, the invariance of the laws of physics toward time displacements, namely, the arbitrary establishment of zero time on a clock, implies the symmetry between past and future, and in the framework of classical mechanics this generates the general energy conservation law. Moreover, just as the existence of a particular law makes it possible to predict an event on the basis of other known events, so the existence of a general conservation law makes it possible to predict new particular laws starting from other known ones. This is one more way to express the internal coherence among the different laws of physics.

The general laws of conservation are quite far-reaching, and they truly form the framework of several natural sciences. For instance, chemistry can be self-defined as the ensemble of events fulfilling the condition that the number of conservation laws must equal the number of chemical elements. This shows that the field of validity of many general invariance laws must be limited and self-restricting because, for example, in the case of the chemical elements this conservation no longer holds true when the events we are considering display energy changes that greatly exceed thermal energy, such as radioactive disintegration that transforms one element into another. But despite their own self-limitations, these general principles have led to the conviction that nature does possess order, and that this order may be understood. This conviction probably first originated in the discovery of the regularities observed in the motion of the heavenly bodies, a discovery which has no doubt influenced the development of all thinking ever since. Today, we are so accustomed to seeing this physical order in the nature that surrounds us that we are no longer surprised by, say, the space order of a crystalline structure, or the time order of a pendulum's motion, or the space–time order of a travelling wave. Furthermore, when speaking of disorder, the example of "molecular chaos" thought to exist in the molecular velocities of an ordinary gas is sometimes mentioned, but actually this is not correct, because if the gas is at thermodynamic equilibrium, there is always a statistical law that governs the distribution probability of the molecular velocities. If no such distribution law existed, or if it could be arbitrarily varied from time to time, the disorder in question would be far greater indeed!

B.3 Biological Order

The living organism looks like a highly correlated system of structures involved in processes, or of processes involving structures—a system which maintains its internal coherence despite the continuous exchange of products with the outer world. Thus biological order is a *dual order of structures and processes,* complementary aspects of the same unit that reciprocally condition each other. Indeed, if taken separately, neither structures nor processes are typical of the biological world, but they certainly fall within the realm of physics. There are highly complex synthetic macromolecules, just as there are complicated, self-controlled processes in technology as well. The unique feature displayed by the biological world is that this highly correlated system generates other similar systems by means of a sequence of processes and structures that keep the species' characters constant despite the variability of each member of the species. Organisms multiply in a favorable environment, and if we limit our observations to organisms, life seems to be a steady growth of ordered matter. But we also know that if we consider the closed physical system composed of the organisms and their environment, then we see that this growth can occur only because of concurrent energy dissipation, and thus because of a growing disorder in the closed system.

Most likely, the idea of biological order came from the observation of some common characters that persist despite the continuous transformation of individuals belonging to a group, something which has made it possible to establish the abstract concept of the species as a specific invariant property of that group, as an ordering criterion when dealing with populations formed by different groups. Mendel's laws governing the genetic behavior of the species are statistical laws that enable us to predict the frequency with which certain hereditary characters will appear, and thus they are a macroscopic expression of this order. Only recently has *this order been stated more precisely, in microscopic terms, as information* contained in the sequence of some particular chemical groups of the DNA macromolecule. This information controls the reproductive process of the structures, which takes place identically in every known form of life. As we saw in Chapter Five, the steady functional order that makes this reproduction process possible can be formally described as a correlated statistical ensemble of significant events. But this functional order differs from physical order, because in physics it is actually possible to fully describe the hierarchy of correlations among events, whereas in biology this is made impossible by the great complexity of the cell, which is already a system of irreducible complexity.

In Chapter Five we saw that both spatially ordered structures and thermal noise contribute to the formation of those significant events that make up functional order. We saw that chance, too, makes a fundamental contribution to the nature of the living. In biology, the term *disorder* is sometimes used to indicate random mistaken reading of some letters in the genetic code, caused by the rare chance that some DNA group may be misidentified. This chance leads to genetic mutations, which all together make it possible for the species to evolve slowly toward the best fitness to its environment. But this evolution is also accompanied by a growing complexity of the organism, and thus it seems incorrect to use the term disorder to express just one side of the process of evolution. This is a continuous process that also displays a growth of order, because of the need to give rise to an internal coherence in a biological entity of ever-increasing complexity.

So if we now limit our observation to an organism set in a partly unfavorable environment, we shall witness the growth of ordered matter as the same characteristic of life we observed above in a favorable environment in the absence of mutations. But if the environment is too unfavorable, we know that life is extinguished forever. For this reason, too, it would seem wise to keep biological order separate from physical order, because life seems to include events that take place within their own limits, which are still unknown.

B.4 The Search for Scientific Order

The idea of order in the natural sciences is thus the result of a continuous effort based upon objective knowledge, a process intended to simplify reality and at the same time to widen its own limits. This objective knowledge can be communicated, and thus recognition of this order among the members of the human community is of obvious social value.

Human beings have a spontaneous desire to look for order when perceiving the world around them. It is probable that this desire to detect the existence of order originated in primitive man's absolute need to establish cause-and-effect relationships in his natural environment, in order to survive. And like the other primary self-preservation mechanisms, this persistent need must also have rooted itself among man's instincts, so that we derive pleasure from the security of being able to predict external events and thus dominate them. Traces of this mechanism can be found in some of the primitive myths. The rhythmically repeated ritual, the fixed reference point of the totem, the seasonal feast for which the entire tribe gathered together—these are all social phenomena that may also be seen as a desire to feel the existence of temporal or spatial order, as expressed in symbolic forms. Religious thinking, which certainly comes from this mythical world, has gradually absorbed and transformed this initial instinct, offering transcendence as a permanent reference point in a world considered ephemeral and confused. Thus the ideal of universal order is born. Dante felt that the personal experience of this universal order was the very purpose for every human life, as he wrote at the end of his *Paradise:*

> In that abyss I saw how love held bound
> into one volume all the leaves whose flight
> is scattered through the universe around.*

Thus Western man has developed that personal drive to find order among observable natural events even when using scientific methods. This is what

*English version by Dorothy L. Sayers (Penguin, London, 1977)

Einstein meant in 1936 when he so admirably stated, "Man tries to create all that can be best adapted to a simplified and comprehensible picture of the world, and somehow he tries to overrule the world of true experience by replacing it with this cosmos of his own." But let us not forget that science considers every instant in time to be identical to any other instant in time, and any point in space to be equal to every other point in space. It is astounding to discover how much of reality can be understood by this outlook. Still, our senses pick up other events, events that cannot be arranged within such a simple measurable framework, thanks merely to the yardstick and the clock. For these events, the signs are the most appropriate expressions, and so the correlation among these signs can be perceived as a sense of immeasurable order, for example, in the unity of a work of art.

LIST OF BOXES

BIBLIOGRAPHY

[1] Arecchi, F. T. 1977. *Interaction of Radiation with Condensed Matter.* Vol. I. IAEA.

[2] Atkins, K. R. 1959. *Liquid Helium.* Cambridge Univ. Press.

[3] Bertman, B., and Guyer, R. A. August 1967. Solid Helium. *Scientific American.*

[4] Born, M. 1962. *Atomic Physics.* Glasgow: Blackie & Son.

[5] Brewer, D. F. 1966. *Quantum Fluids.* Amsterdam: North-Holland Publ.

[6] Burgess, R. E., editor. 1965. *Fluctuation Phenomena in Solids.* New York: Academic Press.

[7] Careri, G., editor. 1963. *Liquid Helium.* XXI Corso Varenna. New York: Academic Press.

[8] ________. Fasella, P., and Gratton, E. 1979. Enzyme Dynamics. The Statistical Physics Approach. *Annual Review of Biophysics and Bioengineering,* **8**, 67.

[9] Decker, A. J. 1965. *Solid State Physics.* Englewood Cliffs, N.J.: Prentice-Hall.

[10] Digiorgio, V. October 1976. The Laser Instability. *Physics Today.*

[11] Domb, C., and Green, M. S., editors. *Phase Transitions and Critical Phenomena.* Several volumes. New York: Academic Press.

[12] Fermi, E. 1957. *Thermodynamics.* New York: Dover Publ.

[13] Fisher, M. E. 1972. Phase Transitions, Symmetry, and Dimensionality. *Essay in Physics*, **4**, 43.

[14] Glansdorff, P., and Prigogine, I. 1971. *Thermodynamic Theory of Structure, Stability, and Fluctuations.* New York: Wiley & Sons.

[15] Goodstein, D. L. 1975. *States of Matter.* Englewood Cliffs, N.J.: Prentice-Hall.

[16] Gorter, C. J., editor. 1955 fol. *Progress in Low-Temperature Physics.* Several volumes. Amsterdam: North-Holland Publ.

[17] Graham, R. 1975. *Fluctuations, Instabilities, and Phase Transitions.* Edited by T. Riste. New York: Plenum.

[18] Green, M. S., editor. 1971. *Critical Phenomena.* LI Corso Varenna. New York: Academic Press.
[19] Haken, H. 1975. Cooperative Phenomena in Systems Far from Thermal Equilibrium and in Nonphysical Systems. *Review of Modern Physics*, **47**, 67.
[20] ________. 1977. *Introduction to Synergetics. Nonequilibrium Phase Transitions and Self-organization in Physics, Chemistry, and Biology.* Berlin: Springer.
[21] Hess, B.; Goldbeter, A.; and Lefever, R. 1978. Temporal, Spatial and Functional Order in Regulated Biochemical and Cellular Systems. *Advances in Chemical Physics*, **38**, 363.
[22] Kittel, C. 1958. *Elementary Statistical Physics.* New York: Wiley & Sons.
[23] ________. 1966. *Introduction to Solid State Physics.* New York: Wiley & Sons.
[24] Koshland, D. K., Jr. October 1973. Protein Shape and Biological Control. *Scientific American.*
[25] Koshland, D. E., and Neet, K. E. 1968. The Catalytic and Regulatory Properties of Enzymes. *Annual Review of Biochemistry*, **37**, 359.
[26] Landau, L. D., and Lifshitz, E. M. 1958. *Statistical Physics.* London: Pergamon Press.
[27] Lehninger, A. L. 1971. *Bioenergetics. The Molecular Basis of Biological Energy Transformations.* 2nd ed. Menlo Park, Calif.: Benjamin/Cummings.
[28] Lwoff, A. 1962. *Biological Order.* Cambridge, Mass.: MIT Press.
[29] Lynton, E. A. 1961. *Superconductivity.* New York: Wiley & Sons.
[30] Ma, S. K. 1976. *Modern Theory of Critical Phenomena.* Reading, Mass.: Benjamin/Cummings.
[31] MacDonald, D. K. C. 1962. *Noise and Fluctuations.* New York: Wiley & Sons.
[32] Mermin, N. D., and Lee, D. M. December 1976. Helium 3 Superfluid. *Scientific American,* **235.**
[33] Nagle, J. F., and Scott, H. L. February 1978. Biomembrane Phase Transitions. *Physics Today.*
[34] Nicolis, G., and Prigogine, I. 1977. *Self-Organization in Nonequilibrium Systems. From Dissipative Structures to Order Through Fluctuations.* New York: Wiley & Sons.
[35] Perutz, M. F. December 1978. Hemoglobin Structure and Respiratory Transfer. *Scientific American,* **239.**
[36] Prigogine, I.; Nicolis, G.; and Babloyantz, A. 1972. Thermodynamics of Evolution. *Physics Today,* **25,** Nov., p. 23, and Dec., p. 38.
[37] Slater, J. C. 1953. *Quantum Theory of Matter.* New York: McGraw-Hill.
[38] Stanley, H. E. 1971. *Introduction to Phase Transitions and Critical Phenomena.* New York: Oxford Univ. Press.
[39] White, R. M., and Geballe, T. H. 1979. Long-Range Order in Solids. *Solid State Physics.* Suppl. 15.
[40] Wilks, J. 1970. *An Introduction to Liquid Helium.* Oxford: Clarendon Press.
[41] ________. *The Third Law of Thermodynamics.* 1961. Oxford Univ. Press.
[42] Wilson, K. G. August 1979. Problems in Physics with Many Scales of Length. *Scientific American*, **241.**
[43] Zemansky, M. W. 1957. *Heat and Thermodynamics.* New York: McGraw-Hill.

INDEX